Applied Probability and Statistics (Continued)

BUSH and MOSTELLER · Stochastic Models for Learning
CHAKRAVARTI, LAHA and ROY · Handbook of Methods of Applied Statistics, Vol. I
CHAKRAVARTI, LAHA and ROY · Handbook of Methods of Applied Statistics, Vol. II
CHERNOFF and MOSES · Elementary Decision Theory
CHEW · Experimental Designs in Industry
CHIANG · Introduction to Stochastic Processes in Biostatistics
CLELLAND, deCANI, BROWN, BURSK, and MURRAY · Basic Statistics with Business Applications
COCHRAN · Sampling Techniques, *Second Edition*
COCHRAN and COX · Experimental Designs, *Second Edition*
COX · Planning of Experiments
COX and MILLER · The Theory of Stochastic Processes
DEMING · Sample Design in Business Research
DODGE and ROMIG · Sampling Inspection Tables, *Second Edition*
DRAPER and SMITH · Applied Regression Analysis
GOLDBERGER · Econometric Theory
GUTTMAN and WILKS · Introductory Engineering Statistics
HALD · Statistical Tables and Formulas
HALD · Statistical Theory with Engineering Applications
HANSEN, HURWITZ, and MADOW · Sample Survey Methods and Theory, Volume I
HOEL · Elementary Statistics, *Second Edition*
JOHNSON and LEONE · Statistics and Experimental Design: In Engineering and the Physical Sciences, Volumes I and II
KEMPTHORNE · An Introduction to Genetic Statistics
MEYER · Symposium on Monte Carlo Methods
PRABHU · Queues and Inventories: A Study of Their Basic Stochastic Processes
SARHAN and GREENBERG · Contributions to Order Statistics
SEAL · Stochastic Theory of a Risk Business
TIPPETT · Technological Applications of Statistics
WILLIAMS · Regression Analysis
WOLD and JURÉEN · Demand Analysis
YOUDEN · Statistical Methods for Chemists

Tracts on Probability and Statistics

BILLINGSLEY · Ergodic Theory and Information
BILLINGSLEY · Convergence of Probability Measures
CRAMÉR and LEADBETTER · Stationary and Related Stochastic Processes
RIORDAN · Combinatorial Identities
TAKÁCS · Combinatorial Methods in the Theory of Stochastic Processes

Evolutionary Operation

A WILEY PUBLICATON IN APPLIED STATISTICS

Evolutionary Operation

A Statistical Method for Process Improvement

GEORGE E. P. BOX
NORMAN R. DRAPER
Department of Statistics
University of Wisconsin

Learning Resources
Brevard Community College
Cocoa, Florida

John Wiley & Sons, Inc.
New York · London · Sydney · Toronto

Library of Congress Catalog Card Number: 68-56159
SBN 471 09305 X
Printed in the United States of America

Preface

This book is about the philosophy and practice of Evolutionary Operation (called EVOP for short), a simple but powerful statistical tool with wide application in industry. Experience has long shown that statistical methods, sometimes quite sophisticated in character, can be of great value in improving the efficiency of laboratory and pilot-plant investigations made by specially trained chemists and engineers. What originally motivated the introduction of EVOP, however, was the idea that the widespread and daily use of *simple* statistical design and analysis during *routine* production by *process operatives* themselves could reap enormous additional rewards.

This text can be used in a number of ways. To learn how to apply simple routine EVOP procedures, only Chapters 1, 4, and 5 need be read. Chapters 2 and 3 provide the basic principles necessary for a sound understanding of EVOP and set out ideas applicable to *any* EVOP scheme. Thus Chapters 1 through 5 constitute a complete EVOP course.

Successful use of Evolutionary Operation by the manufacturer requires education at three levels of the organization:

1. Higher and middle management need to know what it is, to be convinced of its usefulness, to understand how it fits into the organizational structure, and how it relates to other investigational effort.
2. The process superintendent, chemist, and engineer responsible for actual plant operations need sufficient detailed knowledge to enable them to initiate, run, and interpret EVOP programs.
3. The process operator needs to understand the objectives of EVOP and to be trained in this method of operation insofar as it affects him.

How these objectives can be achieved is discussed in Chapter 6.

Chapter 7 briefly discusses a number of techniques related to EVOP and so helps to set it in context. Chapter 8 deals with a variety of questions that often recur in discussions about EVOP. Four appendices expand points mentioned in the text and a comprehensive bibliography is provided.

In this book some time is spent in discussing elementary statistical principles. Experience shows that those who apply these simple principles in the initial running of EVOP schemes will have gone a long way toward understanding the basic principles of good experimental design and analysis and will be ready to study and profit from more advanced statistical applications.

In writing this book the principal audience we had in mind is industrial management—specifically the chemists, engineers, foremen, and process superintendents responsible for running industrial processes. In educational institutions EVOP can be studied profitably by students of management science and chemical and industrial engineering. For the professional statistician the philosophy and special problems of continuous information gathering, analysis, and feedback may be of interest.

We are grateful to a number of people who helped us. An earlier version of the manuscript was reviewed by Joseph J. Brigham, J. Stuart Hunter, Truman L. Koehler, Kenneth D. Kotnour, and Otto A. Kral, all of whom made helpful comments. Raymond E. Niznik helped to prepare the entire final draft of the manuscript and took much of this burden from our shoulders. David E. Tierney performed several calculations and constructed several of the original figures. George M. Minich generated the table of random normal deviates in Appendix 2. Susan D. Anderson typed the earlier drafts and the final version of the manuscript in a characteristically careful and intelligent manner and also constructed several of the original figures. Jin Ju Sheu and Joan L. Lane began the index, which was completed with the help of Nancy A. Draper. Jacobo Sredni made some suggestions at the proof stage, and Agnes M. Herzberg read and commented on the entire set of page proofs.

We are grateful to Frank S. Riordan for his continual encouragement and help in fostering the application of these methods, and to the Monsanto Company for permission to reproduce the "sputnik" photograph. We should also like to thank Richard Bingham, Mavis Carroll, Edward Coleman, Ralph DeBusk, Bruce Drew, William Ellis, Edwin Harrington, Robert Herbert, Richard Freund, William Hunter, William Peters, Maynard Renner, Harry Smith, and particularly Harry Hehner and Truman Koehler for help in various other ways.

Certain portions of this book were prepared initially under the partial support of three sponsors. Research was supported by the Air Force Office of Scientific Research, Office of Aerospace Research, United States Air Force (AFOSR Grant No. AF-AFOSR-1158-66 and contract AF 49(638)-1608), by the United States Navy through the Office of Naval Research

(Contract Nonr-1202(17), Project NR 042-222) and by the Wisconsin Alumni Research Foundation through the University Research Committee (who provided computer time). We gratefully acknowledge this support and also the help of a number of publishers, editors, and authors who gave us permission to reproduce portions of their various publications. Sources are given where the material appears.

Finally, we appreciate the excellent job done by the printers and by the staff of John Wiley and Sons. Marcia Heim and Beatrice Shube were especially helpful.

The final responsibility for errors is, of course, ours; we should be glad to have any that have slipped by us brought to our attention.

GEORGE E. P. BOX
NORMAN R. DRAPER

Madison, Wisconsin
November, 1968

Contents

CHAPTER PAGE

1. THE BASIC IDEAS

1.1 Introduction 3
1.2 Small-Scale and Plant-Scale Investigation 6
1.3 Analogy with Biological Natural Selection 8
1.4 Static and Evolutionary Operation 10
1.5 Cutting the Grass 11
1.6 An Example 13
1.7 An Analysis of the Information Board 15
1.8 A Three-Variable Scheme 16
1.9 The EVOP Committee 18
1.10 When Not to Stop 20

2. SIMPLE STATISTICAL PRINCIPLES ON WHICH EVOP IS BASED

2.1 Industrial Processes, Observations and the Dot Diagram . . . 23
2.2 Frequency Distributions 30
2.3 Distribution Characteristics: Mean and Variance 33
2.4 The Normal Distribution 36
2.5 Estimates of Mean and Standard Deviation from a Sample. . . 41
2.6 Distribution of the Sample Average (or Sample Mean) 46
2.7 Mean Value and Variance of Important Contrasts 50
2.8 Making Inferences in the Presence of Uncertainty: Significance Tests and Confidence Intervals 55

3. THE 2^2 AND 2^3 FACTORIAL DESIGNS

3.1 Factorial Designs 63
3.2 The 2^2 Factorial Design 65

CHAPTER PAGE

3.3 The 2^3 Factorial Design 79
3.4 Dividing the 2^3 Factorial Design into Two Blocks 99
3.5 Summary 104

4. WORKSHEETS FOR TWO-VARIABLE EVOP PROGRAMS

4.1 Introduction 105
4.2 Worksheets for a 2^2 Factorial Design with Added Reference Conditions 109
4.3 Worksheets for a 2^2 Factorial without Additional Reference Conditions 117

5. WORKSHEETS FOR THREE-VARIABLE EVOP PROGRAMS

5.1 Introduction 119
5.2 Worksheets for a 2^3 Factorial Arranged in Two Blocks with a Reference Run in Each Block 120
5.3 Worksheets for a Blocked 2^3 Factorial without Additional Reference Runs 129
5.4 Simplicity and Sophistication 132
5.5 Block-to-Block Variation and the Standard Error of the Phase Mean 132

6. SOME ASPECTS OF THE ORGANIZATION OF EVOLUTIONARY OPERATION

6.1 Training Program 135
6.2 Simulation of Two-Variable EVOP: The EVOP Game 141
6.3 Aids to Successful EVOP 147

7. EVOP, OPTIMIZATION, AND VARIATIONS OF EVOP

7.1 Introduction 152
7.2 Optimization Methods and EVOP 160
7.3 Some Optimization Techniques Related to EVOP 164
7.4 Some Optimization Techniques Unrelated to EVOP 175
7.5 Some Suggested Modifications of EVOP 175

8. COMMENTS AND QUESTIONS ON EVOP

8.1 A Discussion of Some Objections, Comments, and Queries . . . 180

APPENDIX 1. THE APPROXIMATE METHOD OF ESTIMATING THE STANDARD DEVIATION IN EVOP (See Section 4.2) 196

CHAPTER PAGE

APPENDIX 2. GENERATING DATA FOR THE EVOP GAME (See Section 6.2) . . 198

APPENDIX 3. OPTIMAL EMPIRICAL FEEDBACK (See Section 7.1) 200

3A. Choice of Strategy Using Empirical Feedback to Achieve the Greatest Gain in a Fixed Time 200
3B. Choice of Strategy Using Empirical Feedback to Achieve Maximum Increase in a Response y When the Process Function is a Linear Function of a Variable w 209

APPENDIX 4. HOW MANY CYCLES ARE NECESSARY TO DETECT EFFECTS OF REASONABLE SIZE? (See Section 7.1) 211

TABLES

Table I. Normal distribution (single-sided) 218
Table II. Probability points of the normal distribution (single-sided) . . 220
Table III. Probability points of the normal distribution (double-sided) . . 220
Table IV. A table of factors w_k which convert the range of a normal sample of size k into an estimate of the standard deviation . . 222
Table V. A table of values of $f_{k,n}$ 222
Table VI. Useful factors for EVOP calculations 223
Table VII. A short table of random normal deviates 224

REFERENCES AND BIBLIOGRAPHY 227

INDEX 233

Evolutionary Operation

CHAPTER 1

The Basic Ideas

1.1. INTRODUCTION

A typical industrial process passes through many stages of development. Thus in the evolution of a chemical process to produce a certain product, first comes the idea for a promising manufacturing route, followed by (often lengthy) laboratory work to explore its possibilities. The laboratory results provide a preliminary estimate of feasibility, permit realistic objectives to be defined, and may lead to the tentative outline of an industrial process. This outline may then be used to build a pilot plant. The pilot plant is a partway stage between lab and full scale and will have sufficient flexibility to allow rather drastic modifications to be tested. Engineering expertise may now be applied to design a full-scale plant. At each stage an economic analysis is made to determine whether further effort is justified or whether the project shall be abandoned and effort diverted to more promising channels. Assuming that the plant is built then, ideally, it will incorporate the best design possible, given the available knowledge and resources. Subsequent major modification is undertaken only with the greatest reluctance. We have built our apparatus and we must learn to use it. However, there is usually a wide choice of operating conditions available. The small-scale work will have provided "ball park" estimates of concentrations, temperatures, pressures, flow rates, agitation speeds, rates of heating, and rates of cooling. Useful though these estimates are, they usually represent only good first guesses in a continuing process of iteration. This fact is recognized in the special attention given to plant startup. A special technical team is usually assigned during this period, since it is realized that major adjustments may be necessary before the process can be made to perform reasonably well (or sometimes to perform at all).

With the startup phase completed, a further stage of iteration has been gone through and, hopefully, the process will produce a salable product. Almost inevitably, however, the product is being manufactured at lower production rates, at lower yield, and at lower quality than the plant is capable of.

The process of "tuning" still remains to be done. That this tuning can produce rich rewards is clear when we consider typical plant history. It is common to find chemical processes 10 years after startup with production rates two or three times those originally thought possible, with major yield increases, and with a product of greatly improved quality. It is this tuning process that we are concerned with in this book.

Although the job of tuning is a very important task it is not a simple one. Suppose we analyze the standard operating procedure for a chemical process of average complexity into individual component instructions much as we break up a mathematical algorithm into separate instructions in programming the computer. We end up with the following. Instruction 1: "Set concentration 1 to level A" . . . . Instruction 17: "Set temperature 5 to level X" . . . . Instruction 37: "Run agitator 3 at level Z" . . . . The number of such individual instructions can easily be 100 or more. Now suppose we read each instruction to the process supervisor and each time asked the question: "Are you sure that the condition given here is in fact best for your full-scale process?" Many times his answer would have to be: "No, I am not sure. This particular choice of setting arises from small-scale work and we are currently using it, but it may not be the best," or sometimes, "This was once checked on the full-scale process and seemed to be the best, but since then we have changed a number of other variables so that we do not now know for certain that this is still best."

In most industrial organizations considerable effort is made further to improve processes after startup. In addition to the recording and examination of routine plant data, further special studies are frequently carried out in the laboratory, on the pilot plant, and also on the full-scale process by the research and development staff. Such investigations, particularly when they employ efficient modern investigatory tools of statistical design and analysis [see, for example, Davies (1956, 1957)], are most valuable and should, of course, continue.

The shortage of technical personnel, however, inevitably limits the amount of special investigation of this kind. What we here discuss is an additional technique that makes only sparing use of technical manpower and complements the special investigations already referred to.

Let us look at the situation from a somewhat different viewpoint. Because of the special technical effort exerted on process improvement, a manufacturer will normally find a steady increase in productivity from his plants. When he manufactures a large number of different products, the

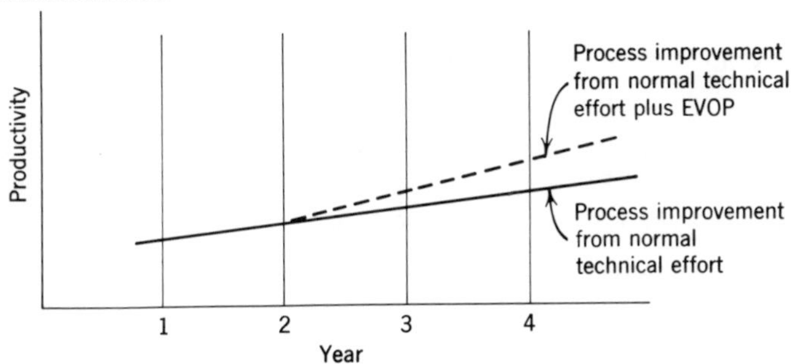

Figure 1.1. Increasing productivity.

average yearly increase is often found to be remarkably constant, as illustrated by the lower sloping line in Figure 1.1.

This is, from one point of view, reassuring; the manufacturer does not know at the beginning of any particular year which particular processes will be improved or what the precise nature of the improvements will be but, nevertheless, he can predict with some exactitude what the over-all effect of applying his technical work force will be. From another point of view the near certainty of the specific amount of process improvement expected during the forthcoming year is depressing. The manufacturer knows that without some new initiative he cannot hope to increase his rate of improvement of productivity.

The purpose of this book is to describe a simple but powerful technique which provides this new initiative. It is called Evolutionary Operation or more briefly, EVOP (pronounced "EVE-OP"). It has been most widely, and successfully applied in the chemical industry, but would undoubtedly be of value in other types of manufacture. Evolutionary Operation is applied on the full-scale plant on a day-to-day basis. It requires no special staff and can be handled by the usual plant personnel after very brief training. Its basic philosophy is that it is inefficient to run an industrial process in such a way that only a product is produced, and that a process should be operated so as to produce not only a product but also *information on how to improve the product.*

Evolutionary Operation is a management tool in which a continuous investigative routine becomes the *basic* mode of operation for the plant and replaces normal static operation. Once everyone connected with the process becomes used to this mode of operation, the question is not "Are we running EVOP?" but "What are we presently investigating with EVOP?"

Evolutionary Operation is not a substitute for fundamental investigation. Such investigation should continue side by side with EVOP. Evolutionary Operation does, however, often indicate areas to which more fundamental approaches could usefully be directed.

1.2. SMALL-SCALE AND PLANT-SCALE INVESTIGATION

As we have remarked, EVOP is a plant-scale technique employed as a basic operating method. Why do we use the full-scale plant for tuning our process and not the laboratory or pilot plant?

1. Because scaleup effects may make "tuning" on other than the full-scale plant pointless.

2. Because, whereas the full-scale runs will be made in any case, pilot and lab runs require additional effort and tie up scarce facilities and technical personnel.

Scaleup

As we have seen, small-scale experimentation can provide invaluable information at the design stage of the process and can furnish first guesses at full-scale operating conditions. However, flow characteristics, heat transfer characteristics, and mixing will all change as the scale of operation is changed and engineering techniques to make allowances for such effects are imperfect.

It is not true, of course, that *completely* different characteristics are to be expected on the large scale. If this were so, design on the basis of small-scale units would be useless. In translating the process from the small scale to the plant scale, the main features of its behavior are usually preserved, but differences in detail can occur which are important economically.

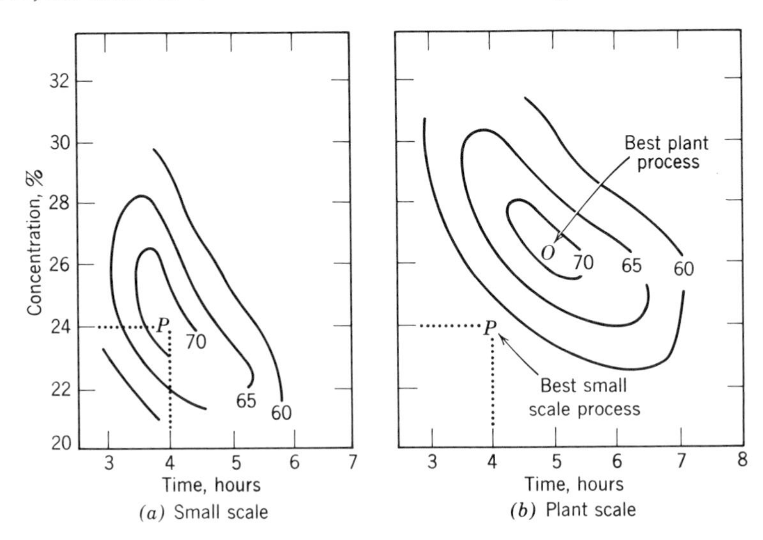

Figure 1.2. *Possible appearance of yield surfaces, showing contours of percentage yield, for a process conducted (a) on the small scale and (b) on the plant scale.*

A simple illustration is given in Figure 1.2. This shows yield contour surfaces for the same chemical reaction conducted on (*a*) the small scale and (*b*) the plant scale. It will be seen that, because of scaleup effects, the yield surface on the plant is distorted and displaced compared with the small-scale yield surface, even though the basic characteristics of the two surfaces are quite similar. The best combination of time and concentration for the small scale (point P) will give disappointingly low yields on the full scale. Efforts to move from the small-scale optimum P to the plant-scale optimum O must, necessarily, be made on the plant itself, since small-scale investigation can only lead back to P. We must climb the *right* hill.

Note. Readers unfamiliar with the representation used in Figure 1.2 may welcome the following explanation. The relationship between a *response*, such as percentage yield, and two variables, such as time and concentration, can be represented by a surface called a *response surface*. In the neighborhood of a maximum response, this surface may have the appearance of a mound, the height of the mound representing the response at some specified conditions of time and concentration. To show the surface in two dimensions, contours of equal yield in the two-dimensional space of time and concentration are drawn just as contours of height are drawn in map making.

The diagram illustrates our point but is, of course, a gross oversimplification of the over-all problem. In practice we have a displacement and a distortion due to scaleup not in a two-dimensional space but in a multidimensional space. Furthermore, in practical situations, our objective would not necessarily be the maximization of yield. We shall return to this point later.

That the manufacturer will need to tune his process is, of course, well recognized. As we have mentioned, some allowances are made by empirical adjustments at startup. Further modification occurs over a considerable period of time, through chance discoveries, new ideas, special experimental investigations and so on. The introduction of EVOP on the plant itself greatly increases the speed at which this progress takes place, and frequently leads to improvements which otherwise would not have been discovered at all.

Conservation of Laboratory and Pilot Facilities

We now return to our second reason for using the full-scale process itself for tuning purposes: even were there no scaleup difficulties, it would usually be wasteful to develop an operating process by small-scale experimentation alone. The cost of such experimentation is high, and to have sufficient laboratory and pilot facilities and technical manpower to carry out process development in this way on all processes simultaneously would be prohibitively wasteful.

In practice such small-scale experimentation is a high-priced luxury. When we commit these facilities and their accompanying specialists to a particular project, we automatically deny them to other projects. By contrast, the full-scale plant itself and the personnel needed to run this plant are already committed in the routine production of product. If these same production runs can be used to provide the needed information, it is obviously inefficient to commit other resources. The full-scale apparatus is available, the men to operate it are there and are trained in its use, the runs are going to be made anyway. Why not use them?

Finally it may be remarked that improvements in plant operation will often result from modifications which cannot be simulated at all on the small scale.

1.3. ANALOGY WITH BIOLOGICAL NATURAL SELECTION

With EVOP we institute a set of rules for *normal plant operation* so that (without serious danger of loss through manufacture of unsatisfactory material) an evolutionary force is at work which steadily and automatically moves the process toward its optimum conditions if it is not operating there already. Such a technique will gradually nudge the operating procedure into the form ideally suited to the particular piece of equipment which happens to be available. To see how this should be done, it is instructive to compare the evolution of industrial processes and the evolution of biological species. Living things advance by two mechanisms:

1. Genetic variability due to various agencies such as mutation.
2. Natural selection.

Chemical processes advance similarly. Discovery of a new route for manufacture corresponds to a mutation. Adjustment of the process variables to their best levels, once the route is agreed, involves a process of natural selection in which unpromising combinations of the levels of process variables are neglected in favor of promising ones.

Figure 1.3 illustrates diagrammatically the possible evolution of a species of lobster. It is supposed that a particular mutation produces a type of lobster with "length of claws" and "pressure attainable between claws" corresponding to the point P on the diagram and that, in a given environment, the contours of "percentage surviving long enough to reproduce" are like those shown in the figure. The dots around P indicate offspring produced by the initial type of lobster. Since those in the direction of the arrow have the greatest chance of survival, over a period of time the scatter of points representing succeeding generations of lobsters will automatically move up the survival surface. This automatic process of natural

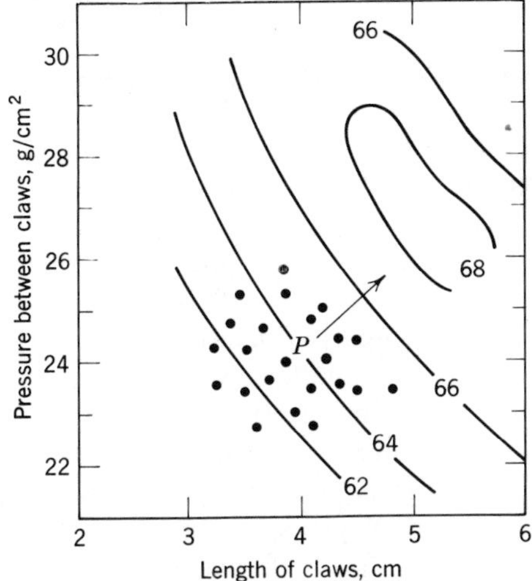

Figure 1.3. *Evolution of a species of lobster. Contours show the percentage surviving long enough to reproduce in a given environment.*

selection ensures, without any special effort on the part of the lobsters, that optimum-type lobsters exist. It also ensures that, if the environment alters so that the survival surface changes, the lobsters will change correspondingly to the new point of maximum survival. We shall try to imitate this process.

It is important to notice that there are *two* essential components of the process of natural selection*:

1. Variation.
2. Selection of "favorable" variants.

* An amusing aspect of one such natural evolutionary process was reported in the London Sunday Express of October 29, 1967. Under the heading "Supercats Have the Catchers Licked," the following appeared:

"Cat-trapping experts of the Royal Society for the Prevention of Cruelty to Animals (R.S.P.C.A.) admitted failure yesterday in their tussle with the supercats of the British Museum. They're just too clever to catch.

The supercats—a semiwild breed who live in the Museum's maze of basements—appear to have developed superior intelligence and avoid every trap.

The R.S.P.C.A.'s Mr. Mike Chester was called in by Museum officials worried by the multiplying cat population in their storage basements and heating system. But he said yesterday: 'I've had to give up and call a truce.

'With most colonies of this kind you can catch the kittens at least. But not the Museum cats. They may never have seen a trap before, but once they spot a wire cage with a pilchard in it they turn the other way.

'The Museum staff see cats all over the basement every day, but when I arrive the cats

1.4. STATIC AND EVOLUTIONARY OPERATION

Routine production is normally conducted by running the plant at rigidly defined operating conditions called the *works process*. The works process embodies the best conditions of operation known at the time. The manufacturing procedure, in which the plant operator aims always to reproduce exactly this same set of conditions, will be called the method of *static operation*. Although this method of operation, if strictly adhered to, clearly precludes the possibility of evolutionary development, yet the *objectives* which it sets out to achieve are nevertheless essential to successful manufacture, for in practice we are interested not only in the productivity of the process but also in the physical properties of the product which is manufactured. These physical properties might fall outside specification limits if arbitrary deviations from the works process were allowed. Our modified method of operation must therefore include safeguards which make the risk of producing appreciable amounts of material of unsatisfactory quality acceptably small.

In the EVOP method a carefully planned cycle of minor variants on the works process is agreed upon. The routine of plant operation then consists of running each of the variants in turn and continually repeating the cycle. The cycle of variants follows a simple pattern, the persistent repetition of which allows evidence concerning the yield and physical properties of the product in the immediate vicinity of the works process to accumulate during routine manufacture. In this way we use routine manufacture to generate not only the product we require but also the information we need to improve it.

Controlled *variation* having thus been introduced into the manufacture, the effect of *selection* is introduced by arranging for the results to be continuously presented to the process superintendent in a way that is easily comprehended. This allows him to see what changes ought to be made to improve manufacture. The stream of information concerning the products from the various manufacturing conditions is summarized on an *information board* which is prominently displayed. This board is continuously brought up to date by a person to whom the duty is specifically assigned. The information is set out in such a way that the process superintendent can at any time see what weight of evidence exists for moving the center of the scheme of variants to some new point, what types of change are undesirable from the standpoint of producing material of inferior quality, how much the scheme is costing to run, and so on.

---

seem to know and vanish. It's selective breeding that does it. Only the cautious, intelligent animals have survived previous trapping attempts. They breed cautious, intelligent kittens.'

Some estimates put the number of cats in the basement at 150."

In making a permanent change in the routine of plant operation, the situation is very different from that which we meet in running specialized experiments. The latter will last a limited time, during which special facilities can be made available. Furthermore, some manufacture of substandard material is to be expected and will be budgeted for. Evolutionary Operation, however, is virtually a permanent method of running the plant and can, therefore, demand few special facilities and concessions. For this reason only changes in the levels of the variables can be permitted whose effects are virtually undetectable in individual runs, and only techniques simple enough to be run continuously by works personnel themselves under actual conditions of manufacture can be employed.

1.5. CUTTING THE GRASS

We have already mentioned that small changes in the process, the effects of which are virtually undetectable in individual runs, are used to provide information for improvement. It will not be immediately obvious how this can be done, so we now enlarge on this point.

Suppose a new plant has just been built. In Figure 1.4 we imagine the latent improvements which could arise from suitable adjustments to the new process measured on some scale such as profitability and arranged in descending order of magnitude. A few very large effects to the left of the diagram will be quickly detected after startup. Their causes will then be tracked down and adjustments and improvement of the process will result.

The reason that these effects can be rapidly detected, hence, exploited,

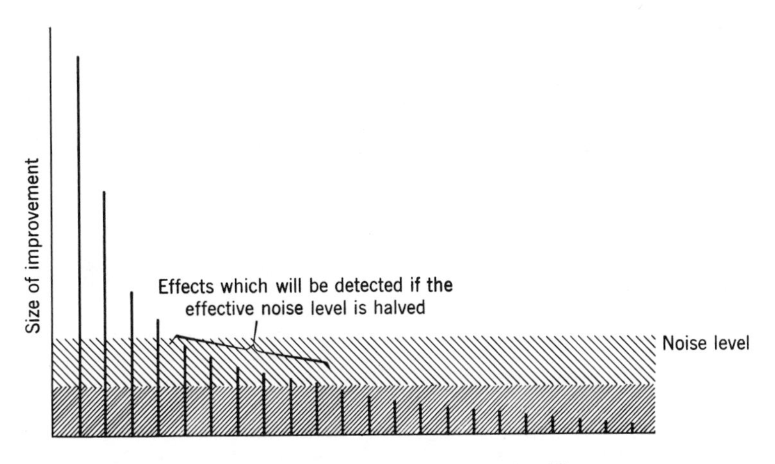

Figure 1.4. Effects and the "noise level."

is that the "signal" they produce is large compared with the *underlying level of variation* of the process, sometimes called the *noise level.*

Noise arises from a variety of sources such as variability of raw materials, inability to maintain precisely input variables at set levels, and from instrument and measurement error. The final variation in the measured output is a composite of all these. The magnitude of the variation is measured by a quantity called the *standard deviation* or *root-mean-square deviation.* The precise definition and method of calculation of the standard deviation, usually denoted by σ (the Greek letter sigma), is described in Chapter 2.

At any given time in the life of a plant the size of the variation, as measured by the standard deviation σ, has some definite magnitude that depends on the degree of process control achieved up to that time. Some specified multiple of this standard deviation can be called the noise level or "the height of the grass." If a change in level of an input variable produces an *effect* in the response which greatly exceeds the noise level, it "sticks out of the grass" and is detected and exploited easily. However, if the effect produced is much smaller than the noise level, it is unlikely to be detected and cannot be exploited.

To discover effects buried in noise we must improve the signal-to-noise ratio. We must either decrease the effective noise level or increase the signal level. In EVOP we do both. The signal is increased by *deliberately* introducing changes of a *carefully chosen kind* in the variables under study. The effective noise level is reduced by repetition of the changes and averaging of the results. The diagram shows how a number of effects previously hidden will be shown up if we cut the noise to one half of its previous value.

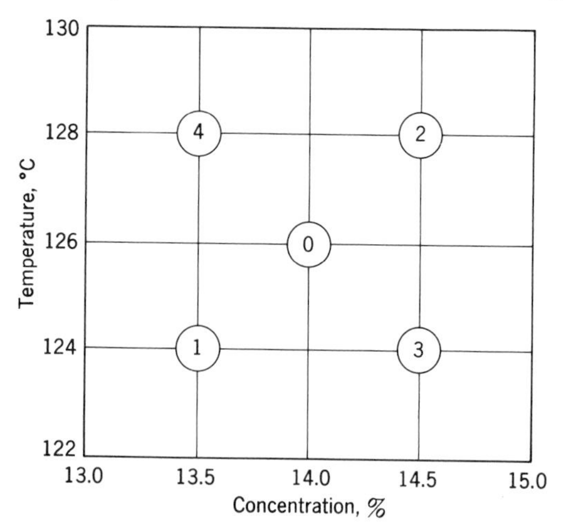

Figure 1.5. *A cycle of variants about the works process.*

1.6. AN EXAMPLE

It is our intention in this chapter to provide a general "bird's eye view" of the whole EVOP picture and then to describe the various elements of the method in greater detail in later chapters. Here we briefly describe a simple EVOP scheme for two variables, postponing more complete details of the design of the scheme and the analysis of the results until later.

The Variables Under Study

At a particular stage of development of a certain batch process, two process variables or factors were being studied. These were the percentage concentration of a certain feed material and the temperature at which the reaction was conducted.

A Pattern of Variants or Design

The scheme of variants is shown in Figure 1.5. The current works process is labeled 0 and the four variants are labeled 1, 2, 3, and 4. One batch of product was made at each set of conditions, which were run successively in the order 0, 1, 2, 3, 4; 0, 1, 2, 3, 4; and so on.

Responses of Interest

Three responses were recorded:

1. The cost of manufacturing unit weight of product. This was obtained by dividing "the cost of running at the specified conditions" by "the observed weight yield at those conditions." It was desired to bring this cost to the smallest value possible, subject to restrictions listed in 2 and 3 below.
2. The percentage of a certain impurity. This should not exceed 0.5.
3. A measure of fluidity. This should lie between the limits of 55 and 80.

The Information Board

The information was recorded by writing in chalk on an ordinary blackboard. Alternatively, wax pencil on a white plastic board or magnetic letters and numbers on a steel board have been used. The essential thing is that it should be a simple matter to erase or remove one number and replace it by another. The scheme set out in Figure 1.6 is not the only one which could have been adopted, but is intended to show a layout of the results that has been found useful in practice.

The *phase* number at the top left-hand corner of the board indicates that two previous phases of EVOP have already been completed on this

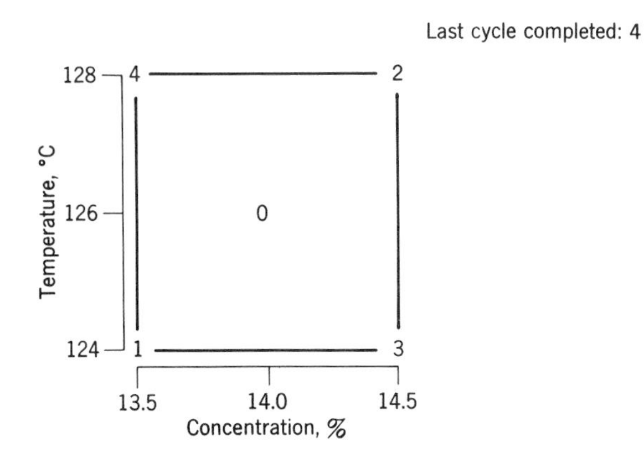

	Cost		Impurity, %		Fluidity	
Requirement	Minimum		Less than 0.50		Between 55 and 80	
	32.6	33.9	0.29	0.35	73.2	76.2
Running averages	32.8		0.27		71.3	
	32.3	33.4	0.17	0.19	60.2	67.6
Two standard error limits	±0.7		±0.03		±1.1	

Figure 1.6. *A portion of the information board after four cycles.*

process. In general these might have involved other variables or the same variables at other levels. In order that the new results may be considered in proper relation to those obtained previously, the final average values recorded in previous phases should also be available (for example, on sheets of paper pinned to the board). The *cycle* number in the top right-hand corner indicates that four cycles of this third phase of operation have been completed. There follows a plan of the cycle of variants being run. The table below summarizes the current situation. First are shown the requirements that must be satisfied. These are followed by the running (i.e., up-to-date) averages at the various manufacturing conditions set out so as to follow the plan of the cycle of variants. This arrangement makes it easy to appreciate the general implications of the results.

Error Limits

A measure of the reliability (or dependability) of the individual running averages is supplied by the *two standard error* limits (2 S.E. limits). As we explain in more detail later, at any given stage of the investigation these limits indicate a range which almost certainly includes the true value. Thus, for example, the average of the fluidity measurements made at a percent concentration of 14.5 and a temperature of 128°C is 76.2. This value is subject to experimental error, but we know from the error limits that the true fluidity at these conditions is almost certainly included in the range of 76.2 ± 1.1. As each cycle is completed, the calculated error limits become narrower and narrower, reflecting the fact that the running averages are becoming more and more reliable. The sizes of these error limits at any given stage inform the process superintendent to what extent the apparent differences in performance which he sees may be mere random fluctuations and to what extent they probably represent real effects.

1.7. AN ANALYSIS OF THE INFORMATION BOARD

To make a full analysis of the situation shown in Figure 1.6 would require the introduction of ideas discussed in later chapters. Our immediate purpose is to show only the attitude with which such an analysis should be made. This we can achieve by using a somewhat inadequate appraisal based only on ideas which are familiar to the beginner.

After studying the information board and combining the ideas it conveys with production requirements and his special knowledge of the process, the superintendent can make one of two basic decisions.

1. To allow the scheme to proceed unchanged and wait for additional information from the next cycle.
2. To modify operation in some way, so beginning another phase of the EVOP scheme.

Under alternative 2, a number of possibilities are available. Some of these are as follows:

a. Adopt one of the variants as the new "works process" and recommence the cycle about this new center point.

b. Explore an indicated favorable direction of advance and recommence the cycle about the best conditions found. (This exploration may be done, for example, by making a series of tentative advances in the indicated direction, at each stage running the new conditions and the previously best conditions alternately.)

c. Substitute new variables for one or more of the old variables. There will normally be a large number of variables waiting to be tested on the EVOP scheme.

d. Modify the pattern of variants in any way which seems desirable from the information available at the time. In particular, when after a few cycles no appreciable effects are found, it may be decided to vary the present variables again but over somewhat wider ranges.

What decision was, in fact, made in the example that provided the data in Figure 1.6? From the running averages we can see that a reduction of the percentage concentration apparently reduces cost, decreases percentage impurity, and reduces fluidity. If temperature is decreased, the effect on cost is small but apparently favorable, and a marked reduction in impurity occurs. However, decrease in temperature appears to be also accompanied by major reductions in fluidity. Now fluidity must be maintained above 55. It looks as though this limit is rapidly reached if *temperature* is much reduced, but that a fairly large reduction in *concentration* could probably be made before this limit was approached. Although a large reduction in impurity which could be achieved by a temperature decrease would be welcome, it is not essential to meet the specification at this time. For these reasons it was decided, in fact, to reduce concentration *alone*. In phase 4 the three sets of conditions (concentration, temperature)—(13%, 126°), (13.5%, 126°) and (14%, 126°)—were compared by running several cycles at these points. The first point gave rise to a mean cost of 32.1 (a useful reduction), with an impurity level of 0.25 and a fluidity of 60.7 (still well above 55). This point (13%, 126°) then was selected as a starting point for further investigation. It was designated as the new "works process" and became a new base for further cycles of the evolutionary process.

The Art of the EVOP Game

We can seek from this example that the actions to be taken in various circumstances are not *precisely* specified. Human judgment is a very important part of an EVOP investigation as it should be in any investigation. Automatic rules which would in effect take the responsibility for the running of his process out of the hands of the process superintendent are bad both psychologically and practically. A properly run EVOP scheme ensures that the process superintendent is continuously fed with clear information about his process on which he then takes what action he deems appropriate. In applying this information he should bring into play the attitudes used in playing, say, chess. Although there are certain principles in this game, judgment is called for at every stage to evaluate the possibilities of future success or failure and to act accordingly.

1.8. A THREE-VARIABLE SCHEME

Our previous example showed a typical EVOP scheme for examining *two* variables and provided a rough idea of how the results would be examined.

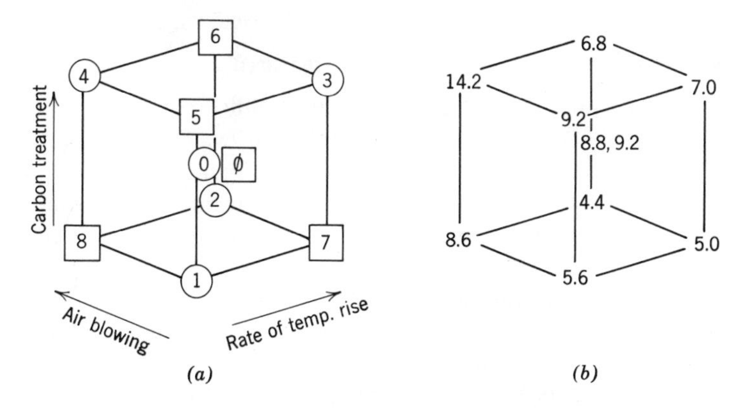

Figure 1.7. *Display of pattern of variants and results for a three variable EVOP scheme.* (*a*) *Cycle of variants.* (*b*) *Results for mean texture after five cycles.*

We now briefly describe a typical three-variable scheme. As before, a geometrical picture is helpful in understanding the situation. Figure 1.7*a* shows a cycle of variants used to examine the effects of changing the three variables:

1. Rate of temperature rise.
2. Air blowing.
3. Carbon treatment.

The duplicated center point (runs 0 and Ø) was the standard works process. The variants 1, 2, 3, 4, 5, 6, 7, 8, were the eight variants obtained by combining the high and the low level of each of the variables in all possible ways. Figure 1.7*b* shows the average "texture" observed after five cycles of the scheme. Given that the object was to reduce this measure of texture to as small a value as possible, the reader may find it interesting to consider the conclusions he would reach from a consideration of these values.

So far we have considered simple schemes in which two or three variables were simultaneously studied. Although more variables can be examined in an EVOP scheme, it is often unwise to do so. The reason is that although, as we discuss later in more detail, it is theoretically possible to improve investigational efficiency by simultaneously considering more variables, in practice two or three variables seem to be about the most that can be handled in a *routine* manner on the full-scale process. If an investigatory scheme on the full-scale process becomes too disruptive, excuses will be found to discontinue it. It is better to settle for a modest scheme that will actually be used and can be lived with than to press for a theoretically better one which will quickly fizzle out.

Responses Which Cannot Be Precisely Measured

The above example serves, incidentally, to illustrate that it is not necessary to confine EVOP only to investigations in which the response can be directly measured. In the case illustrated in Figure 1.7, the response was the "texture" of a particular product—an important if somewhat esoteric quality. To obtain response values, a set of artificial standard samples, judged by experts to have a range of textures in approximately uniform steps, was prepared, and each sample was given a numerical score. Each manufactured sample was then compared with the standards, matched to the standard most like it, and scored accordingly. Many other difficult responses can be assessed in a similar manner. Examples are *caking* (of powdered product) and *fading* (of photographic film under an accelerated test).

1.9. THE EVOP COMMITTEE

In a foregoing section, 1.3, we made a comparison between the natural evolutionary process and EVOP. These processes differ in one vital respect. In nature the variants occur spontaneously, but in our artificial evolutionary process we have to introduce them. Variants involving the levels of temperature, concentration, pressure, etc., are natural choices, but there are many less obvious ways in which manufacturing procedure can be tentatively modified. Frequent instances of marked improvement due to some innovation never previously considered in a process that has been running for many years testify to the existence of valuable modifications waiting to be discovered.

To make our artificial evolutionary process really effective, therefore, one more circumstance is needed—we must set up a situation in which *useful ideas* are continually forthcoming. An atmosphere for the generation of such ideas is perhaps best induced by bringing together, at suitable intervals, a group of people with special, but different, technical backgrounds. In addition to plant personnel themselves, obvious candidates for such a group are, for example, a research man with an intimate knowledge of the chemistry of similar reactions to that being considered and a chemical engineer with special knowledge of the type of plant in question. The intention should be to have complementary rather than common disciplines represented.

These people should form the nucleus of a small EVOP committee, meeting perhaps once a month, whose duty it is to help and advise the process superintendent in the performance of EVOP. The major task of such a group is to discuss the implications of current results and to make suggestions for future phases of operation. Their deliberations will frequently lead to the formulation of theories which, in turn, suggest new leads that can be pursued with profit.

Since questions of modification of certain physical properties of the manufactured product may arise, a representative of the department responsible for the quality of manufacture should also be on the EVOP committee. If a statistician is available as well, more may be obtained from the results and more ambitious techniques can be adopted if and when they are necessary.

Scientific Feedback

The *scientific feedback* occurring in EVOP is perhaps its most important aspect. At regular intervals the process data are examined and the process is discussed by intelligent technical people having a wide range of knowledge and experience in several different areas. The committee members, if they do their job properly, will provide helpful scientific analysis of results as well as a constant flow of new ideas to be incorporated into the investigation. Without this constant stimulus, EVOP may "bog down" and perhaps fail entirely.

We made the point earlier in this chapter that EVOP is run by plant personnel. The use of specialists on the EVOP committee does not seriously

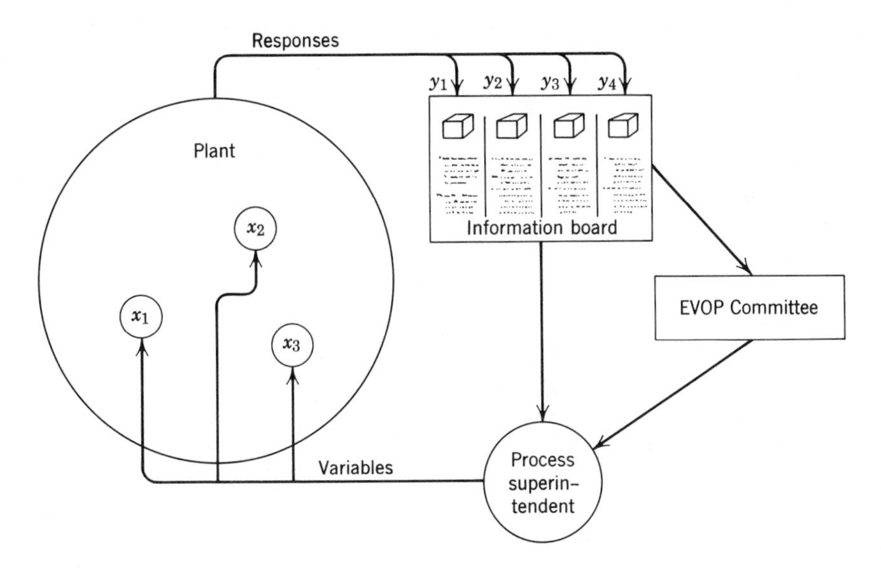

Figure 1.8. Diagrammatic representation of the "closed loop" provided by Evolutionary Operation.

vitiate this principle. In practice, the time spent by the specialists is perhaps one afternoon a month. The ultimate responsibility for running the EVOP scheme still rests with the process superintendent and *not* with the specialists, who serve only in an advisory capacity.

With the establishment of the EVOP committee all the requirements for an efficient evolutionary production routine are satisfied and the *closed loop* illustrated in Figure 1.8 is obtained. We are thus provided with a practical method of process improvement that requires no special equipment and can be applied to almost any manufacturing process, whether the plant concerned is simple or elaborate.

1.10. WHEN NOT TO STOP

With an alert team of workers new ideas should be continually forthcoming and the evolutionary method becomes virtually a permanent mode of operation and should be so regarded. Only if it seemed that more would be lost than gained from the evolutionary procedure would the reintroduction of static operation be justified. In practice it is found that even very small gains will justify the continual operation of the evolutionary method. The situation at any given time can be appraised by the use of a pictorial

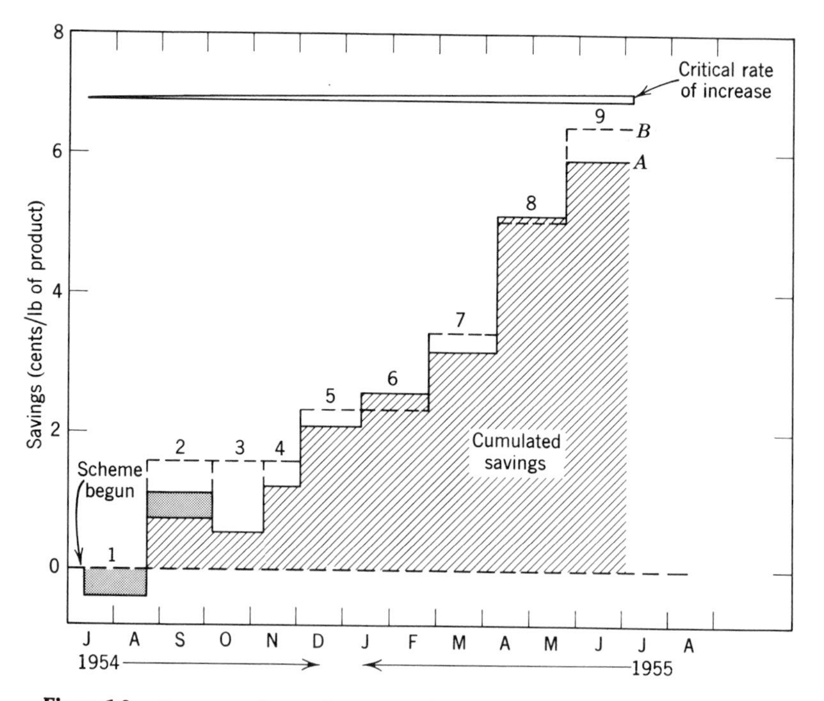

Figure 1.9. Progress of an evolutionary scheme, showing critical rate of increase.

log like that shown in Figure 1.9. The full line A in this diagram shows the savings in cents per pound which have been achieved in the various phases of operation. The dotted line B shows the savings that would have resulted if the center or "works process conditions," appropriate for each particular phase, had been run. On the assumption of constant throughput, the shaded area is proportional to the accumulated savings resulting from the scheme, and each of the rectangular areas between the dotted and full lines shows the accumulated "expenses" of running the scheme during that phase. The darkened area in phase 1 represents the cumulative expenses for the scheme during this phase and is debited from the cumulated savings in phase 2.

Whereas each phase of Evolutionary Operation, and consequently the expenses associated with it, lasts for only a limited time, any improvements which result go on for as long as the process is used. Suppose it is assumed that process improvements will go on earning money for p years after they are discovered and that the running of the evolutionary scheme adds c cents/lb to the cost of the product. Then the question whether at any instant of time Evolutionary Operation should be continued may be resolved by comparing the rate of improvement r, expected to be produced by the evolutionary process (measured in cents per pound per year), with the critical rate of improvement r_0 given by $r_0 = c/p$. For if it is expected that the evolutionary scheme will need to be run for time t years to produce an improvement at the end of that period of rt cents/lb, and if k lb of product is made per year, then the total saving during the p years for which the discovery is used will be $rtkp$ cents. During this time, kt lb of product will be made and the loss due to running the evolutionary scheme will be ckt cents. Thus the scheme will pay off if $rtkp$ is greater than ckt, that is, if r is greater than $r_0 = c/p$.

As an example consider the situation in Figure 1.9. Should the scheme there shown be continued or not? Let us suppose (very conservatively) that improvements on this process are expected to go on earning money for three years after their discovery, so that p is put equal to 3. Suppose also that c is taken to be the average of the values experienced in the nine previous phases. This gives the value $c = 0.3$ cent/lb. We then find for the critical rate $r_0 = 0.3/3 = 0.1$ cent/lb/year. Thus, so long as the rate of improvement due to the evolutionary process is expected to be at least as great as 0.1 cent/lb/year, the evolutionary scheme should be continued. This critical rate of increase is shown diagrammatically at the top of Figure 1.9. It will be seen that the actual rate of improvement which had been experienced over the previous year was about 6 cents/lb/year (about 60 times the critical rate), and there is no evidence as yet of any flagging in this rate of improvement. There is, therefore, no doubt whatever that this scheme should be continued.

Experience gained over a large number of schemes since the inception of EVOP in 1954 has confirmed that the foregoing example is typical and that EVOP should be regarded as virtually a permanent mode of operation.

It is psychologically wrong to talk of EVOP as "experimental manufacture" or to say that EVOP is "experimentation." An experiment is a limited and special program performed for a limited period. EVOP is part of the normal routine *operation* of the plant.

Note: Some of the material of this chapter previously appeared in *Applied Statistics* [see Box (1957)] and is reproduced here by permission of the editors.

CHAPTER 2

Simple Statistical Principles on Which EVOP Is Based

The calculations required in the running of an EVOP scheme are basically simple. They have been systematized and can be conveniently carried through on work sheets described in Chapters 4 and 5. All the information required for the routine conduct of EVOP is in fact contained in Chapters 1, 4, and 5. It is highly desirable, however, that those responsible for running EVOP schemes have some understanding of the elementary statistical principles on which they are based. These principles are set out in this chapter, and Chapter 3 provides a similar basic understanding of the statistical designs most often employed.

2.1. INDUSTRIAL PROCESSES, OBSERVATIONS, AND THE DOT DIAGRAM

A Typical Industrial Operation

Suppose we have an industrial process in which measured quantities of several ingredients are placed in a reactor and are mixed, heated, and allowed to combine chemically to produce a product. Such a process might be conducted either as a batch operation or as a continuous operation. In a batch operation the ingredients would be loaded into the reactor, the reaction would be allowed to take place, and the product would be removed. The reactor would be recharged and the sequence repeated. In a continuous operation the ingredients would be fed continuously into the reactor and the product would be continuously removed. In either case the running of the process at a fixed set of reactor conditions for a fixed period would be called a *run*. For a batch process a run might refer to the manufacture of a single batch or of several consecutive batches made under the same conditions. For a continuous process a run would refer to a fixed period of continuous operation of the reactor under constant conditions. In an EVOP program deliberate changes in process conditions are introduced. In this chapter, however, we shall for the most part be studying the chance variation that can occur in results when *no* deliberate changes in conditions are made; that is, we shall be studying the variation that could occur from run to run owing to chance causes alone. We do this so that we can learn how to distinguish real effects caused by a deliberate change in process conditions from chance effects that might be expected even if no intentional changes in conditions had been introduced.

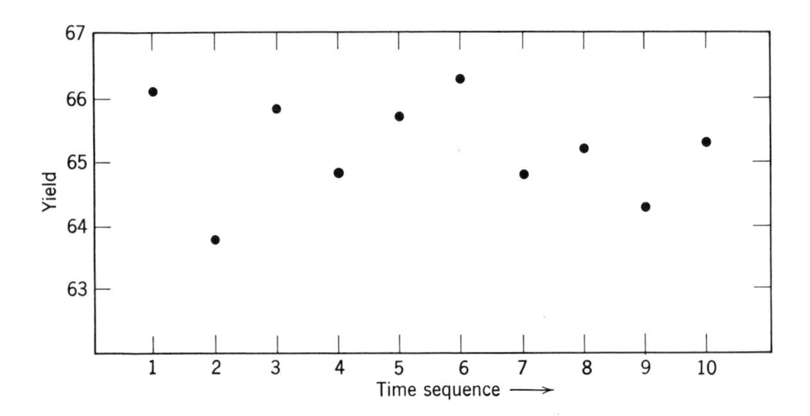

Figure 2.1. *Ten yield observations plotted in time sequence.*

Variation in Results

Even when the process conditions are held constant, nevertheless the observed results vary from run to run. This run-to-run variation arises from many different sources. Measurement errors and analytical errors play their part, but a major cause of variation is the impossibility of reproducing precisely the intended conditions during any particular run.

Plotting Data in Time Order

Suppose that for a particular chemical process, observations of percentage yield for 10 successive runs were as follows: 66.1, 63.7, 65.8, 64.8, 65.7, 66.3, 64.8, 65.2, 64.3, 65.3. These results are plotted in time sequence in Figure 2.1.

Serial Correlation of Observations

The plotting of results in time sequence, as shown in Figure 2.1, is a valuable means of looking at production data. In fact, it is the basis of the well-known quality control chart. Such a plot may reveal a basic trend, a jump in yield level, or the existence of an outlying observation. Such peculiarities may suggest deterioration in lab standards, a change associated with a particular batch of raw material, maloperation of a particular kind, and other "assignable causes" leading to corrective action.

In other cases pseudoperiodic drifting above and below the average may occur. Effects of this kind are particularly common with data from continous processes where disturbances to the system are smoothed out by mixing

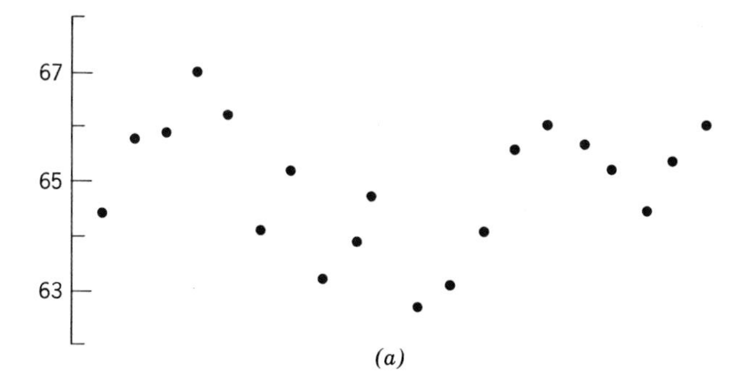

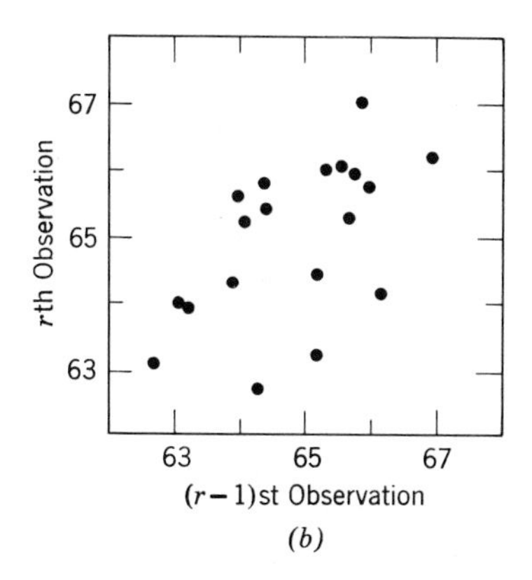

Figure 2.2. *(a) A series of observations exhibiting local positive serial correlation. (b) A lag-1 serial plot for these data.*

and may exhibit themselves in several successive observations. The data plotted in Figure 2.2*a* showing a tendency of this kind are said to exhibit local *positive serial correlation* between successive observations. This means

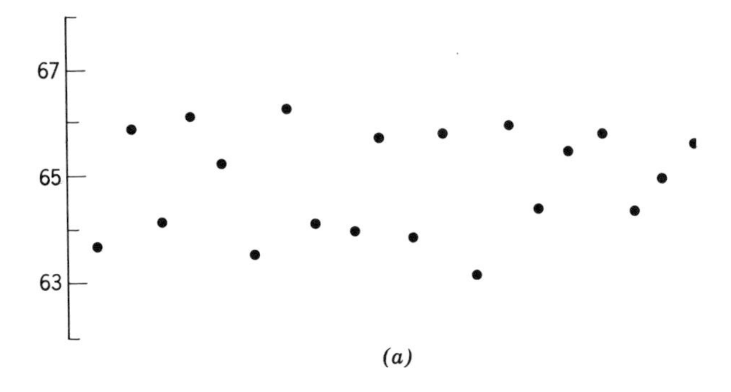

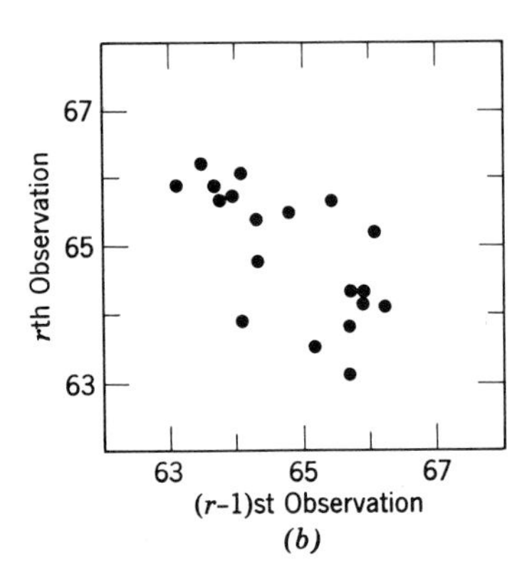

Figure 2.3. *(a) A series of observations exhibiting local negative serial correlation due to carryover. (b) Lag-1 serial plot for these "carryover" data.*

that, if a certain observation has a positive deviation from the average, neighboring observations are likely to deviate positively as well, whereas if a certain observation has a negative deviation, its neighbors are likely to deviate negatively. The correlation between observations one step apart is called *lag* 1 *serial correlation*. It can be demonstrated by plotting each observation (except the first) against the observation preceding it. The *positive* lag 1 serial correlation in the data of Figure 2.2*a* is shown by the "lower left to upper right" tendency of the pattern of points plotted in

Figure 2.2*b*. Similar plots made for observations two steps apart, three steps apart, and so on, might reveal the existence of serial correlation of higher lags.

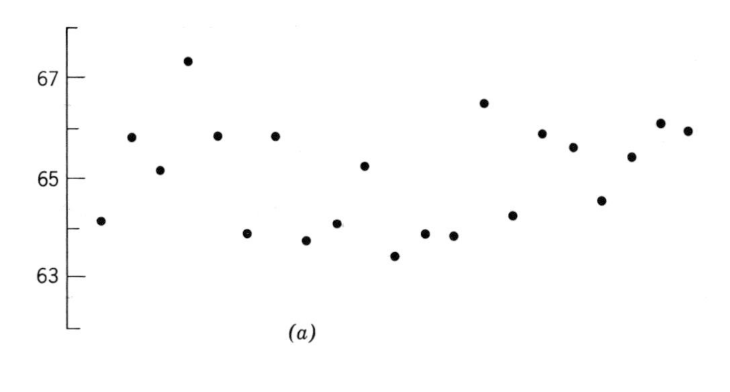

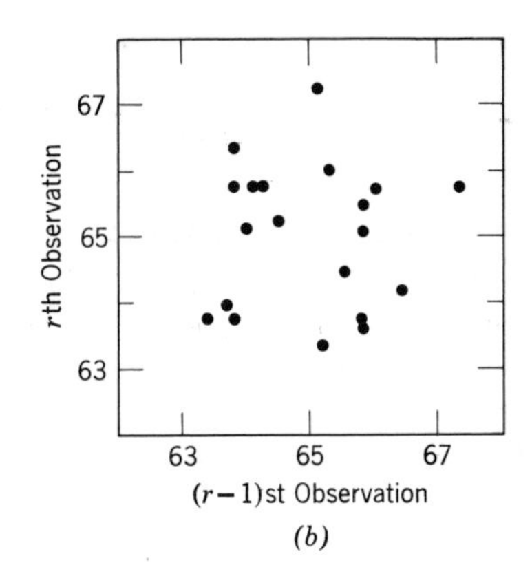

Figure 2.4. *(a) A random series of observations. (b) Lag-1 serial plot for this random series.*

Negative serial correlation between successive observations can also occur. One cause is a phenomenon which occurs in batch processes known as *carryover*. This can happen as follows. Suppose that, for a particular batch, incomplete recovery occurs because some of the product is left in the pipelines and pumps of the reactor system. The recorded yield for this batch will be unusually low. In the next batch, however, there would be a tendency for the material left behind to be recovered, thus giving an unusually high batch yield. A pattern of observations may result like those

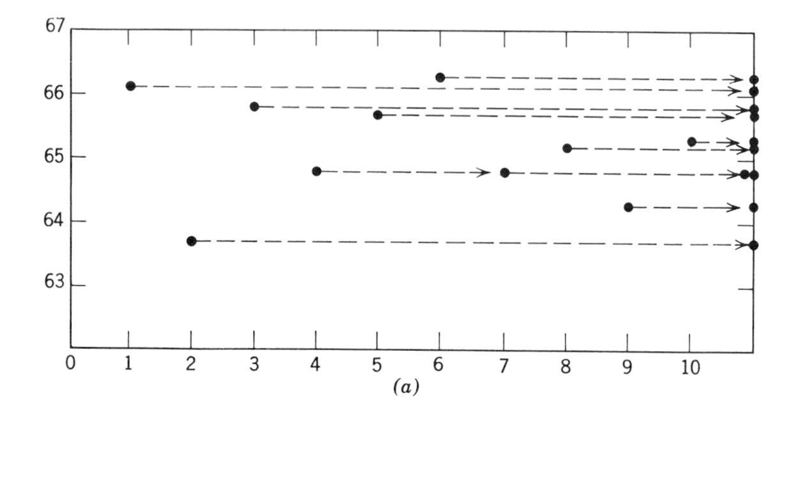

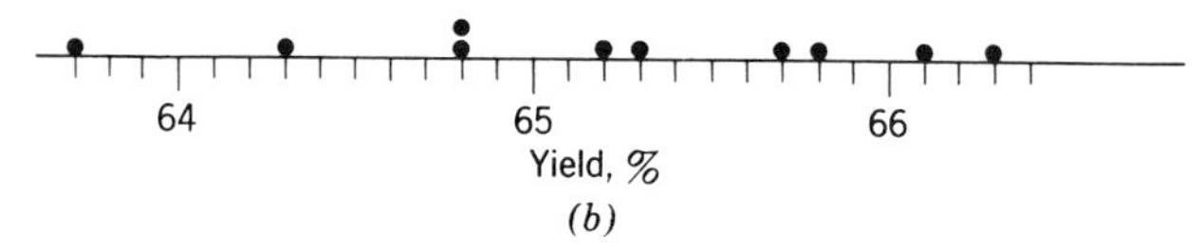

Figure 2.5. *(a) The yield sequence of Figure 2.5 collapsed onto the vertical axis to produce a dot diagram. (b) A dot diagram for percent yield drawn on the horizontal axis.*

plotted in Figure 2.3*a*, in which a high yield tends to be followed by a low one, and vice versa. The existence of negative lag 1 serial correlation for these data is shown by the "lower right to upper left" pattern of Figure 2.3*b*.

The study of serial correlation patterns is one of the techniques used in time-series analysis. Such analysis of plant data can be rewarding. The interested reader is referred to texts such as that of Jenkins and Watts (1968) and Box and Jenkins (1969).

Assumption of Random Variation

For the time being we are going to assume that we have a situation where the results vary *randomly* about some mean value. This assumption means that the probability that a deviation from the mean level will exceed a given size is completely unaffected by the value taken by any other observation. When this is true, we say that the deviations are statistically independent. In particular, statistical independence implies that there is no serial correlation of any kind between observations. Figure 2.4*a* shows a piece of a random series. As would be expected, the lag 1 serial plot (Figure 2.4*b*) shows no evidence of correlation. The assumption of randomness, although probably never completely true, will usually provide an adequate approximation for our present purpose. One reason is that it is necessary in EVOP to make comparisons only within sets of observations that are close together in time. In fact, in the schemes we discuss, we shall never make comparisons between observations more than five intervals apart. Departures from randomness such as commonly occur usually have much less influence on comparisons of this kind than on comparisons between widely separated observations. We shall further discuss certain implications of the random assumption later.

The Dot Diagram

When it can be assumed that observations vary independently, the time sequence of the observations is irrelevant and, consequently, nothing is lost by ignoring it. In particular, on the basis of this assumption, nothing would be lost by collapsing the observations of yield plotted in Figure 2.1 onto the vertical axis, as shown in Figure 2.5*a*, to provide what is usually called a *dot diagram*. This is more conveniently drawn on a horizontal axis, as shown in Figure 2.5*b*, rather than on a vertical axis.

The dot diagram is a valuable device for displaying the distribution of a small body of data (say up to 20 observations). In particular it calls to attention:

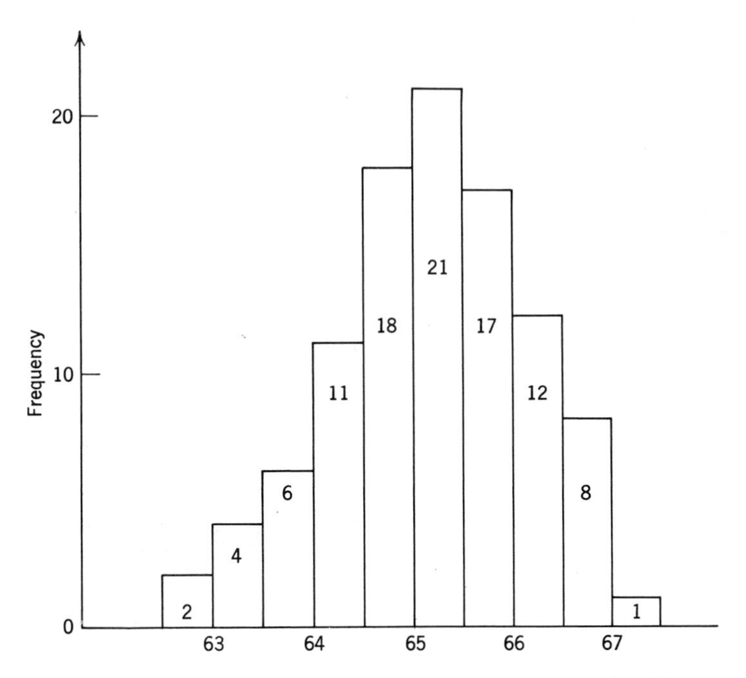

Figure 2.6. Frequency distribution for 100 observations of yield.

1. The general *location* of the observations. (In the example we can see that the yields are clustered near the 65% level rather than, say, 85 or 35%.)

2. The *spread* of the observations. (In the example they range over about three percentage units.)

2.2. FREQUENCY DISTRIBUTIONS

When a large body of data is being studied, the individual dots in a dot diagram become confused. We are better able to appreciate the data by

Table 2.1. Frequency distribution of 100 observations

Class Interval	Number of Observations in Interval = Frequency, f [1]
Below 62.5	0
62.5–63.0	2
63.0–63.5	4
63.5–64.0	6
64.0–64.5	11
64.5–65.0	18
65.0–65.5	21
65.5–66.0	17
66.0–66.5	12
66.5–67.0	8
67.0–67.5	1
Above 67.5	0
Total	100

[1] Some convention must be adopted for observations which fall on the boundary of an interval. It is usually simplest to count these observations as adding one-half to each of the neighboring groupings.

constructing a frequency distribution. To do this, we can mark off the axis in equal intervals of some chosen size and pile the dots which fall in each interval at the central value of the interval. Alternatively on each interval we can construct a rectangle whose area and thus, for equal intervals, whose height is proportional to the number of observations in that interval. Between 10 and 20 intervals covering the relevant range are usually adequate. Figure 2.6 shows a frequency distribution for 100 observations of yield from a production process. The 100 values are distributed in the selected intervals as shown in Table 2.1.

The number of observations falling into a class is called the *class frequency*. Thus, in this example, two observations fell in the first class, which has interval 62.5–63.0, so that the frequency for this class is 2.

A frequency distribution like Figure 2.6 provides a striking and informative visual impression of the distribution of the data. In particular, it shows the location and spread of the observational values.

If we were working with a very large number of observations, we could afford to make the frequency classes narrower and more numerous and still maintain a reasonable number of observations in each class. If the total number of observations were very large, the class intervals could be extremely small and the shape of the diagram would usually approach a smooth curve like that shown in Figure 2.7.

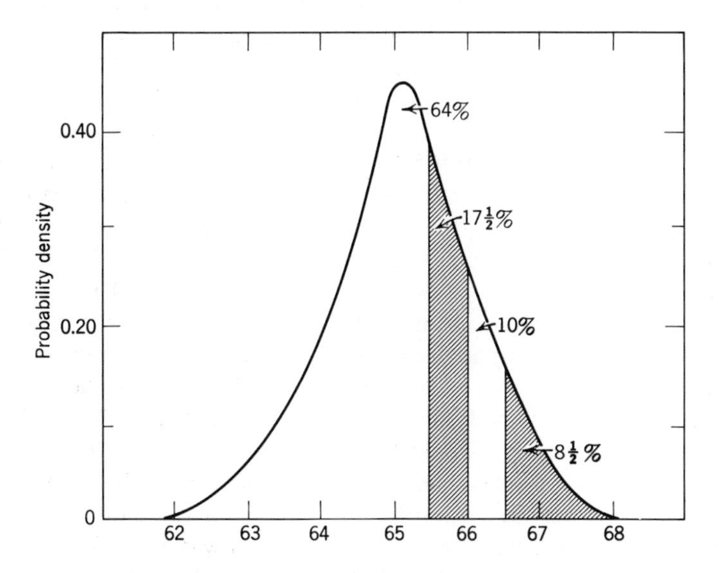

Figure. 2.7 ***A probability distribution for percent yield.***

If we arrange the vertical scale so that the area under this curve is unity, the curve is called a *probability* distribution and its height is said to measure the *probability density*. The name *probability distribution* is often shortened and we may talk simply of the *distribution* of some characteristic such as yield or purity.

The Concepts of Population and Sample

The large (conceptually infinite) collection of all the results that *might* occur as the result of carrying out a particular operation is called the *population.* A probability distribution such as that of Figure 2.7 describes this population, telling us the relative frequencies with which results of different kinds can occur.

In most cases the complete population of possible results is not actually available. We do not know the population. Nevertheless, the *idea* of an underlying population of possible results from any given operation is a very important concept. When we make a run and determine yield at a particular set of reactor conditions, it is as if we were taking part in a celestial lottery. We can imagine an enormous lottery drum containing an infinity of tickets. Each ticket is marked with a value of percentage yield that might be obtained when we try to run the apparatus at the chosen conditions. The probability distribution describes the relative numbers of tickets in the drum inscribed with the various yield results. If, as we are supposing, the operation in which we are engaged is one in which the results are statistically independent, no available information can tell us which particular kind of result we will get. The operation culminating in the yield result is thus simulated by the *random* drawing of a ticket from the drum. If we knew the probability distribution for the population concerned, then, although we could not know in any particular case what *specific* result we would get, we could make probability statements about possible results. For example, the vertical scale in Figure 2.7 has been arranged so that the total area under the curve equals one. The area under the curve to the right of any given yield value will thus represent the *proportion* of yields in the total population which exceed this value; for example, the area to the right of 66.5 (shaded in Figure 2.7) is $8\frac{1}{2}\%$ of the total area under the curve. This would imply, in our analogy, that of the total number of lottery tickets, $8\frac{1}{2}\%$ would have numbers higher than 66.5. On the assumption that the yield values are varying randomly about the mean, this would imply that the *probability* of obtaining a value of yield greater than 66.5 at this particular fixed set of conditions was $8\frac{1}{2}\%$. Similarly, the shaded area between 65.5 and 66 is $17\frac{1}{2}\%$ of the total area under the curve in Figure 2.7, so that the probability of obtaining a yield between 65.5 and 66 is $17\frac{1}{2}\%$.

Now suppose that all that we have are the 10 observations plotted in Figure 2.5. From a statistical point of view we would regard them as a *random sample* from a (hypothetical) infinite population described by an (unknown) probability distribution; that is to say, we could think of our 10 results as if they had been a *sample* of 10 ticket drawings from the celestial lottery drum. Now, although all that we actually have is the sample of 10 observations from the process, we should like to be able to make statements about what the process will do *in general;* that is, we should like to be able to make statements about the *population.* One principal object of

Statistics is to argue from the particular to the general—that is, from the sample (which *is* available but is not of much interest in itself) to the population (which is conceptual and *not* available but is of vital interest).

2.3. DISTRIBUTION CHARACTERISTICS: MEAN AND VARIANCE

To discuss the hypothetical population and its distribution function, we have to have ways of measuring its characteristics. A characteristic of a distribution of particular importance is one which measures general position or location. In Figure 2.8 we see two distributions of identical shape but of different location; one is located near the value 40, the other near the value 60.

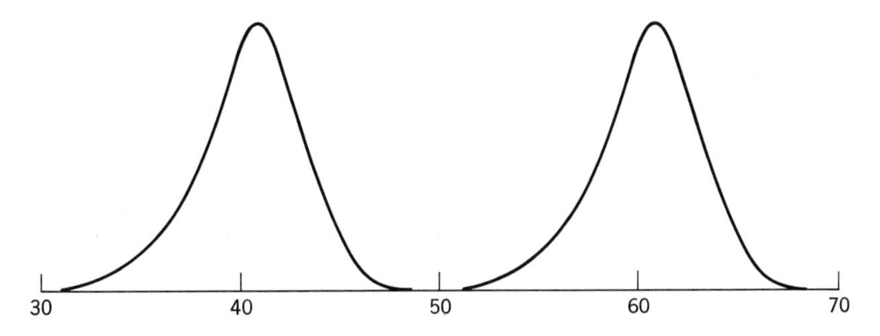

Figure 2.8. Two distributions having different location.

The Mean μ: A Measure of Location

The most useful measure of location for our purpose is the *mean* of the distribution. For a population which contains an infinite number of observations, the mean is usually defined in terms of an integral. We avoid this complication here by supposing that the population contains N observations and that N is very large. We suppose that $y_1, \ldots, y_N$ are the N observations comprising the population and we denote the mean by μ (the Greek letter mu). So that

$$\mu = \frac{y_1 + y_2 + \cdots + y_N}{N} = \sum_{i=1}^{N} \frac{y_i}{N}.$$

The mean μ locates physically that vertical axis about which the distribution would balance.

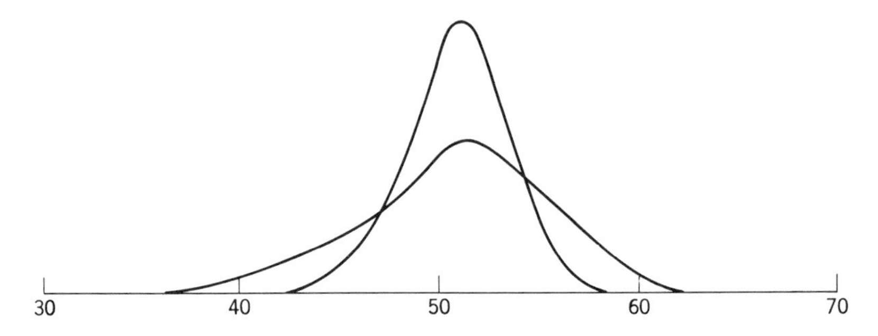

Figure 2.9. *Two distributions having different spread.*

The Standard Deviation σ: A Measure of Spread

The mean μ supplies information about location, but other kinds of information about the distribution are important also; for instance, if we were told that the mean yield of a process were 50%, it could imply that a result of 50% could be expected every time or, alternatively, that half the time we would get 100% and half the time nothing! We will certainly be helped to understand what the population in question is like if, in addition to its mean, we are also supplied with some measure of its spread. In Figure 2.9 we see two distributions that have the same location, but one is more widely spread than the other.

As a measure of spread we employ the *mean of the squared deviations about the mean:*

$$\sigma^2 = \frac{(y_1 - \mu)^2 + (y_2 - \mu)^2 + \cdots + (y_N - \mu)^2}{N} = \sum_{i=1}^{N} \frac{(y_i - \mu)^2}{N}.$$

This is called the *variance* of the population. The square root of this quantity, which has the same units as the observations themselves, is called the *standard deviation* and, as we have indicated already, is denoted by the Greek letter sigma. Thus

$$\sigma = \left[\sum_{i=1}^{N} \frac{(y_i - \mu)^2}{N}\right]^{1/2}.$$

The mean and variance of a distribution give us information on the important characteristics of location and spread respectively. They do not, however, completely characterize the distribution. To make precise statements of probability it is necessary to know something about the shape of the distribution and, if it were true that any shape of distribution were possible, our task would be a difficult one. It is fortunate that, in the situation in which we very often find ourselves in industrial experiments, we can approximately predict the shape of the distribution which will usually occur. We now discuss this point more fully.

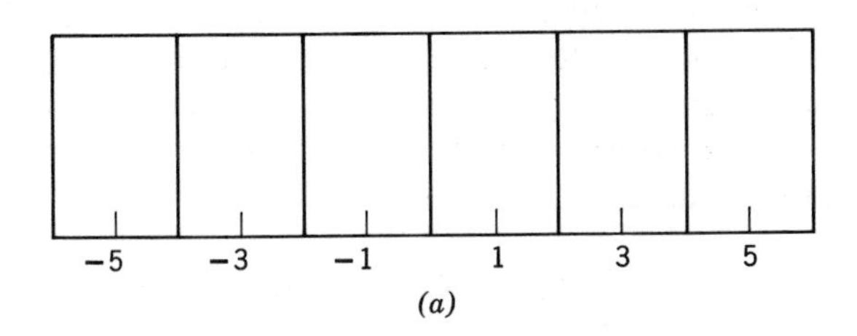

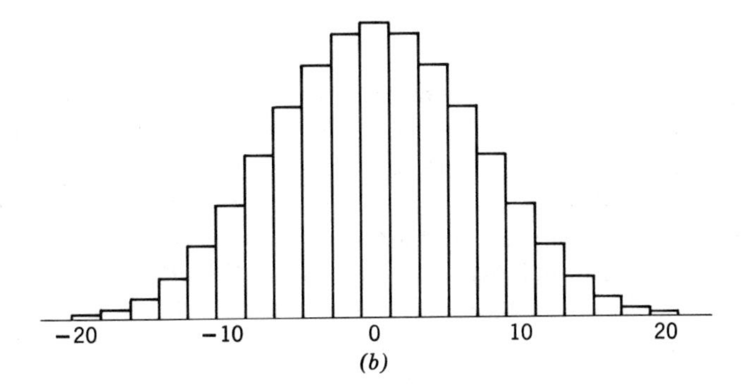

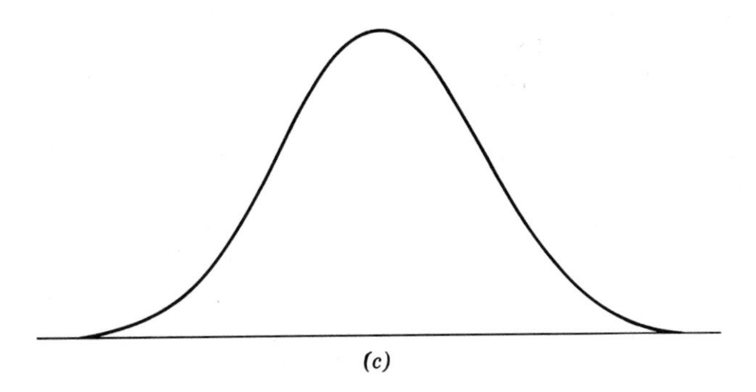

Figure 2.10. *The central limit effect and the normal distribution:* (*a*) *distribution of error* e_1 *from one source;* (*b*) *distribution of error* $e = e_1 + e_2 + e_3 + e_4$ *from four independent sources* (*c*) *normal distribution.*

2.4. THE NORMAL DISTRIBUTION

As we remarked in Section 2.1, observations on an industrial process inevitably vary. The deviations $e = y - \mu$ from the true mean μ can arise from a multiplicity of sources including, for example, analytical error, variation in the quality of raw materials and additives, errors in measuring the product quantities and flow rates, leaks in the piping system, effects from a changing ambient temperature, errors in measuring the weights of the final product, and so on. The over-all "error" e that occurs in a particular observation is necessarily some function

$$e = f(e_1, e_2, \ldots, e_m)$$

of component errors which we denote by $e_1, e_2, \ldots, e_m$. If the component errors vary over fairly small ranges, this can be written approximately as

$$e = k_0 + k_1 e_1 + k_2 e_2 + \cdots + k_m e_m,$$

where the k's are constants. There is a theorem of mathematical statistics, called the *central limit theorem*, which says roughly speaking * that, provided the influence of the various sources of error are of comparable magnitude, the distribution of e tends to the *normal distribution*, as the number of components is increased, no matter what the distributions of individual e's may be. The normal distribution is such that the logarithm of the probability density associated with a given result falls off as the square of the discrepancy between that result and the true mean value. For illustration, a normal distribution is shown in Figure 2.10*c*. The distribution has a number of properties that we discuss shortly.

An example of the central limit effect is illustrated by Figure 2.10. Suppose a single component of error e_1 could take values $-5, -3, -1, 1, 3, 5$ with equal probability. Then the occurrence of random errors from this source could be simulated by the throw of a die with $-5, -3, -1, 1, 3, 5$ inscribed on its six faces. The distribution of error from this single source would then be represented by Figure 2.10*a*. The distribution of aggregate error, $e = e_1 + e_2 + e_3 + e_4$ from four equally important components e_1, e_2, e_3, e_4, each having this kind of distribution, would be simulated by the total scores from four such dice and is shown in Figure 2.10*b*. It will be noticed that the distribution is very similar in shape to the normal distribution shown for comparison in Figure 2.10*c*. The more error components included, the more the distribution of their sum would resemble the normal

* A rigorous statement of the theorem includes a number of conditions for its validity. These conditions will normally be satisfied in the circumstances of our applications.

distribution. In practice, the distribution of error e will not be *exactly* normal. Nevertheless the normal distribution usually provides an approximation which is adequate for our purpose.

Any normal distribution is completely specified by two measures, the mean μ and the standard deviation σ. Figure 2.11 illustrates this point. From Figure 2.11*a* it will be seen that the normal distribution is symmetric about its mean μ which measures its location. It will also be seen how the

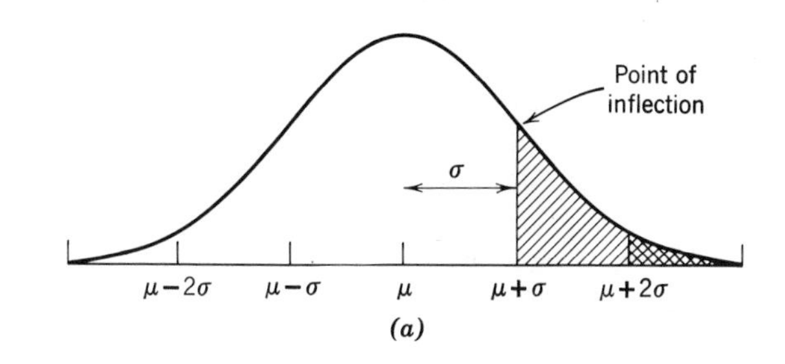

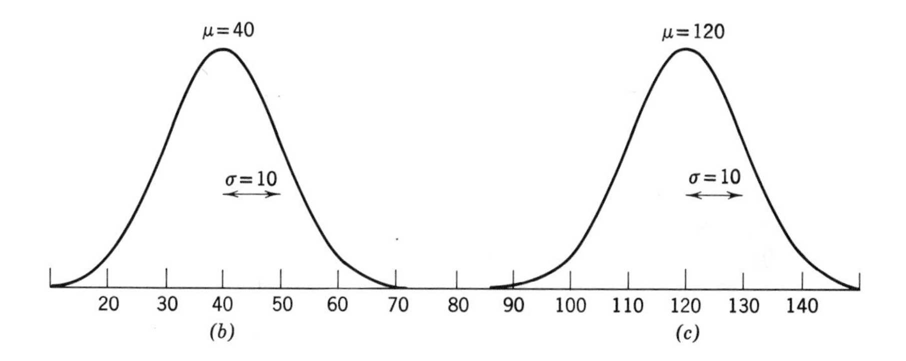

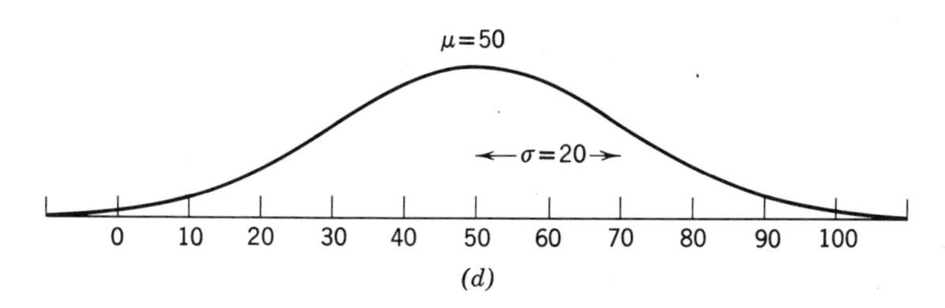

Figure 2.11. *The normal distribution.*

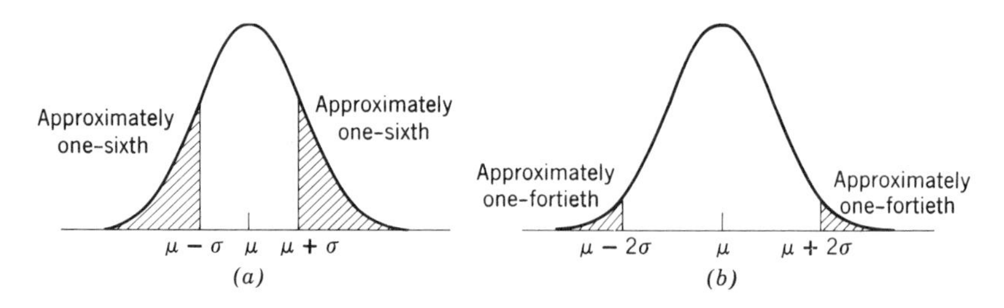

Figure 2.12. *Areas under the normal distribution function.*

standard deviation σ measures the spread of this distribution. Actually σ measures the distance from the center of the distribution to what is called the *point of inflexion* of the curve. The point of inflexion is the point at which the curve changes from convex to concave. Figures 2.11*b*, 2.11*c*, and 2.11*d* show normal distributions with mean and standard deviations respectively given by $\mu = 40, \sigma = 10$; $\mu = 120, \sigma = 10$; $\mu = 50, \sigma = 20$. The distributions all have the same characteristic "normal" shape. The parameters μ and σ merely serve to locate and spread each distribution appropriately.

As we have seen, probabilities are measured by *areas* under the probability distribution. The following facts are worth memorizing. For the normal distribution, about one-sixth of the observations in the population will *exceed* the mean by one standard deviation or more. This means that the area under the curve and to the right of the vertical line marked $\mu + \sigma$ in Figure 2.12*a* contains about one-sixth of the total area under the curve.

Another way of saying this is that the probability that any single observation y will exceed the value $\mu + \sigma$ is about one-sixth. Because of the symmetry it is also true that about one-sixth of the observations have a

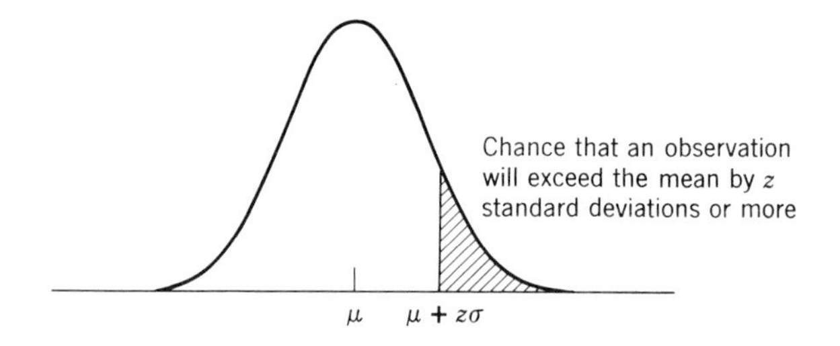

Figure 2.13. *The one-sided probability recorded in Table I (p. 218.)*

value less than $\mu - \sigma$, with parallel statements throughout. Facts like these can be put to immediate and practical use. For example, suppose we know that the per cent yield for the cyclohexane process is approximately randomly and normally distributed with mean $\mu = 63$ and with a standard deviation $\sigma = 3$, and we wish to know the probability of exceeding a yield of 66. Since $66 = 63 + 3 = \mu + \sigma$, the required probability is about $\frac{1}{6}$.

About $\frac{1}{40}$ of the observations in a normal population exceed the mean by two standard deviations or more. This means that the area beneath the curve and to the right of the vertical line marked $\mu + 2\sigma$ in Figure 2.12*b* contains about $\frac{1}{40}$ of the total area under the curve.

Another way of saying this is that the probability that any single observation y will exceed the value $\mu + 2\sigma$ is about $\frac{1}{40}$. Thus for the distributions (*b*), (*c*), and (*d*) in Figure 2.11, the probabilities of exceeding 60, 140, and 90, respectively, are each equal (very nearly) to $\frac{1}{40}$ or 2.5%.

If we consider deviations from the mean in *both* directions, we can say that for the normal distribution about $\frac{1}{3}$ $(= \frac{1}{6} + \frac{1}{6})$ of the observations deviate from the mean *in one direction or the other* by one standard deviation or more and about $\frac{1}{20}$ $(= \frac{1}{40} + \frac{1}{40})$ deviate from the mean *in one direction or the other* by two standard deviations or more. When both directions are considered the probabilities are said to be *two-sided* probabilities; otherwise they are *one-sided.*

Thus, if the question were asked, "What is the probability that the yield for the cyclohexane process previously mentioned will deviate from its mean of 63% by more than 3 percentage units *either way?*" (i.e., be less than 60% or greater than 66%), the answer would be "about one-third" and this is a two-sided probability statement.

We frequently need probabilities associated with deviations other than one or two standard deviations from the mean. Table I (page 218) shows the chance, expressed as a decimal, of an individual observation y departing fro.n the mean *in the positive direction* by z standard deviations or more for various positive values of z. Thus the table records the area shaded in Figure 2.13 and provides a one-sided probability. (If two-sided probabilities are needed, the tabular entries should be doubled.) Notice that when $z = 1$ we obtain $p = 0.1587$ which is "about one-sixth" as stated above; and when $z = 2$ we obtain $p = 0.0228$ which is "about one-fortieth" as stated above.

The table can be used for other purposes; for example, if we need the probability that an observation lies between z_1 standard deviations and z_2 standard deviations above the mean, we want to find the area shaded in Figure 2.14*a*. This area can be found from Table I, p. 218, as the difference between the shaded area in Figure 2.14*b* and the shaded area in Figure

2.14*c*. In this example the deviations are both positive. When both deviations are negative, the same calculation can be performed as if both were positive.

The probability that an observation y lies between $\mu - z_1\sigma$ and $\mu + z_2\sigma$ is obtained by adding areas obtained from the table, as illustrated in Figure 2.15*b* and 2.15*c*. The reader should be warned that there are many tables of the normal distribution probability levels. Some are of the form of our Table I and provide the area *above* the point $\mu + z\sigma$. Other tables provide the area *up* to $\mu + z\sigma$ or the area between μ and $\mu + z\sigma$. Any one of these tables can be used to derive any other by simple addition and subtraction.

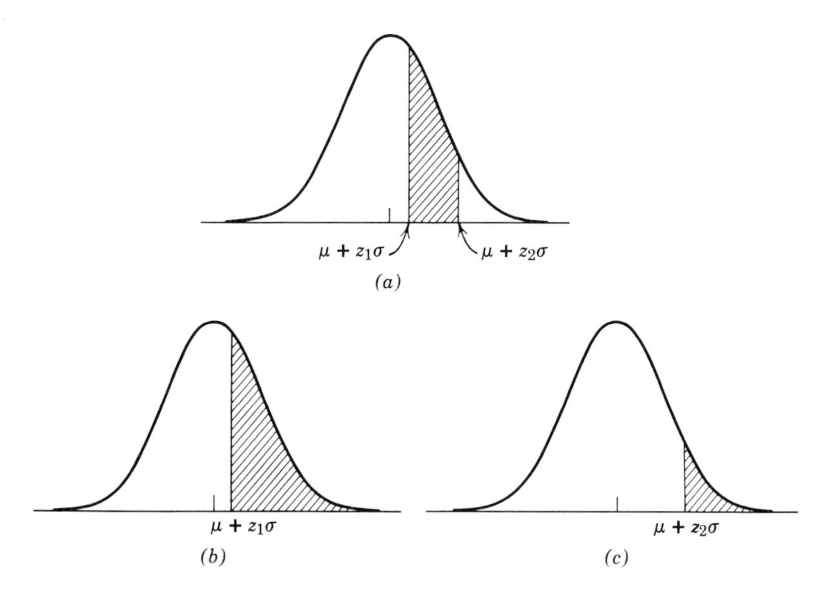

Figure 2.14. *Area (a) as a difference between areas (b) and (c).*

As a further exercise in the use of Table I, consider the following question: If, as before, the percentage yields for the cyclohexane reactor are normally and randomly distributed about a mean of $\mu = 63$ with a standard deviation of 3, what is the chance that the percentage yield from an individual run will be between 66 and 72? The yields of 66 and 72 represent deviations from the mean of $(66 - 63)/3 = 1$ and $(72 - 63)/3 = 3$ standard deviations in the positive direction, i.e., to the right of the mean respectively. From Table I we see that the chance of an observation exceeding one standard deviation ($z = 1$) is 0.1587, whereas the chance of an observation exceeding three standard deviations ($z = 3$) is 0.0013. The probability of an observation lying between 1 and 3 standard deviations from the mean is thus $0.1587 - 0.0013 = 0.1574$. The required probability is thus 15.74%, so that about one in 6.36 runs could be expected to produce yields between 66 and 72.

2.5. ESTIMATES OF MEAN AND STANDARD DEVIATION FROM A SAMPLE

We have seen that to have complete knowledge of a normally distributed population we have to know only its mean μ and its standard deviation σ.

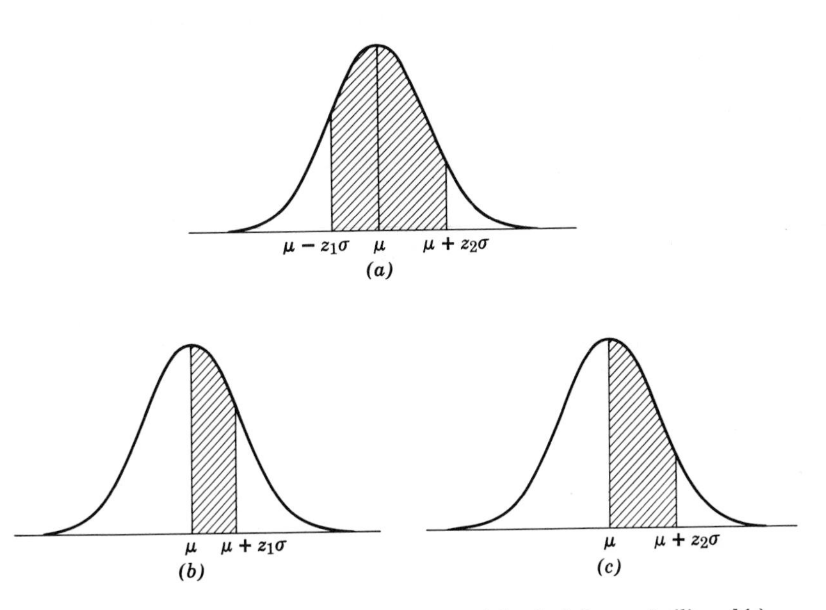

Figure 2.15. *The shaded area in (a) is the sum of the shaded areas in (b) and (c).*

We have also noted that in practice we never have available the complete population of results and so we never actually know μ and σ. What we do have is a *sample* of a few observations from which we would like to make probability statements about the population or equivalently (if it is assumed normal) about μ and σ. We can do this by calculating *estimates* of μ and σ from our limited sample of observations.

Suppose we have a small sample of n observations $y_1, y_2, \ldots, y_n$. Then an estimate of the population mean μ is supplied by the *sample average* (also called the *sample mean*)

$$\bar{y} = \frac{y_1 + y_2 + \cdots + y_n}{n} = \sum_{i=1}^{n} \frac{y_i}{n}. \tag{2.5.1}$$

Furthermore, an estimate of the population variance σ^2 is provided by the *sample variance*

$$s^2 = \sum_{i=1}^{n} \frac{(y_i - \bar{y})^2}{n - 1}. \tag{2.5.2}$$

The *sample standard deviation* is taken to be s, the positive square root of the sample variance.

It is natural to ask: "Why do we divide by $n - 1$ in the formula for s^2 and not by n?" In seeking to understand the formula for the sample variance it will be recalled that the population variance is defined by

$$\sigma^2 = \sum_{=1}^{N} \frac{(y_i - \mu)^2}{N},$$

where N is the very large number of items in the entire population. If we wanted to estimate σ^2 from the sample $y_1, y_2, \ldots, y_n$ and we *knew* the population mean μ, we should use the formula

$$s^2 = \frac{\sum_{i=1}^{n} (y_i - \mu)^2}{n},$$

which exactly matches the formula above for the population variance. Usually, however, we do not know μ so we replace it by $\bar{y}$. Now it can be demonstrated that the quantity

$$\sum_{i=1}^{n} (y_i - a)^2,$$

which is the sum of squares of deviations of the observations about an arbitrary value a, has its smallest possible value when $a = \bar{y}$.

Thus, unless by some fluke, $\bar{y} = \mu$ exactly,

$$\sum_{i=1}^{n} (y_i - \bar{y})^2$$

will be smaller than

$$\sum_{i=1}^{n} (y_i - \mu)^2$$

and can never exceed it, so that, on the average,

$$s^2 = \frac{\Sigma(y_i - \bar{y})^2}{n}$$

would be an underestimate of σ^2. It turns out that this bias is exactly allowed for by replacing n by $n - 1$ as in (2.5.2) and it is now customary to use this divisor.

In many textbooks it is recommended that the sum of squares appearing in the formula for the sample variance be calculated using the "short-cut" formula

$$\sum_{i=1}^{n} (y_i - \bar{y})^2 = \sum_{i=1}^{n} y_i^2 - \frac{\left(\sum_{i=1}^{n} y_i\right)^2}{n}, \tag{2.5.3}$$

where $\sum y_i^2$ is called the *uncorrected sum of squares* and $(\sum y_i)^2/n$ is the *correction for the mean.* The two forms shown are algebraically equivalent; but care must be taken in using the second which is very susceptible to rounding error. The reason is that if the numbers y_i are quite large the quantities $\sum y_i^2$ and $(\sum y_i)^2/n$ will both be large, whereas their difference may be quite small. If we use this form, therefore, we must be careful to retain sufficient significant figures so that the difference is meaningful. Since in practice we would usually wish to examine the individual deviations $y_i - \bar{y}$ (*the residuals*) in any case, direct calculation from the first formula is usually preferable.

Table 2.2. Calculation of $\bar{y}$ and s^2

	y_i	$\bar{y}$	$y_i - \bar{y}$	$(y_i - \bar{y})^2$
	66.1	65.2	0.9	0.81
	63.7	65.2	−1.5	2.25
	65.8	65.2	0.6	0.36
	64.8	65.2	−0.4	0.16
	65.7	65.2	0.5	0.25
	66.3	65.2	1.1	1.21
	64.8	65.2	−0.4	0.16
	65.2	65.2	0.0	0.00
	64.3	65.2	−0.9	0.81
	65.3	65.2	0.1	0.01
Sum	652.0	652.0	0.0	$6.02 = \sum_{i=1}^{10} (y_i - \bar{y})^2$

Calculation of $\bar{y}$ and s. To illustrate the calculation of the sample mean and sample standard deviation we use the data shown in Table 2.2. For the sample mean or sample average we have

$$n = 10, \qquad \sum y_i = 652,$$

so that

$$\bar{y} = \tfrac{652}{10} = 65.2,$$

and for the sample variance,

$$n - 1 = 9, \qquad \Sigma(y_i - \bar{y})^2 = 6.02, \qquad s^2 = \frac{6.02}{9} = 0.67, \qquad s = 0.82.$$

We obtain the same value for s^2 through the second formula above. Since

$$\sum_{i=1}^{10} y_i^2 = 42{,}516.42, \qquad \frac{\Sigma(y_i)^2}{n} = \frac{(652)^2}{10} = 42{,}510.40,$$

we have

$$s^2 = \frac{42{,}516.42 - 42{,}510.4}{9} = \frac{6.02}{9} = 0.67,$$

exactly as before.

A Short-cut Calculation for Estimating the Standard Deviation

It is desirable to simplify EVOP calculations as much as possible. Calculation of $\bar{y}$ is simple enough and, although the calculation for s is not difficult, it is easier and sufficient for our purpose to use a method based on the *range* (the numerical difference between the greatest and least observation in the sample). For example, the range of the sample

$$20, 4, 37, 15, 18, 39$$

is $39 - 4 = 35$. (See Figure 2.16*a*.) When negative numbers occur, full account must be taken of the sign. For example the range of the sample

$$20, -4, 37, -15, 17, -39$$

is $37 - (-39) = 76$. (See Figure 2.16*b*).

On the assumption that observations arise from a normal distribution, we can estimate σ by multiplying the range by a factor w which varies according to the number of observations in the sample. Values of w when the numbers of observations are between 2 and 12 are given in Table 2.3.

We now use this method on the 10 observations of Table 2.2. The largest observation is 66.3, the smallest is 63.7; the range of the sample is thus

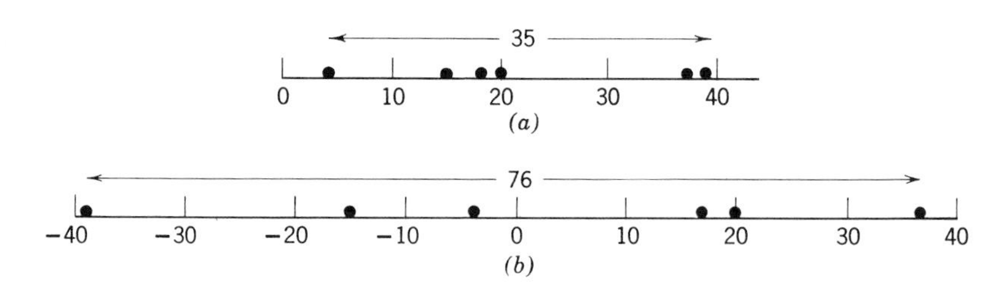

Figure 2.16. *The ranges of two samples.*

Table 2.3. To estimate the standard deviation of a normal distribution, multiply the range of a sample of n observations by the corresponding value of w

n	w[1]
2	0.886
3	0.591
4	0.486
5	0.430
6	0.395
7	0.370
8	0.351
9	0.337
10	0.325
11	0.315
12	0.307

[1] Note that, for $2 \leqq n \leqq 10$, w is approximately equal to $1/\sqrt{n}$. This is useful to remember, since it permits rough calculations to be made when a table of w is not available.

$66.3 - 63.7 = 2.6$. From Table 2.3, $w = 0.325$ when $n = 10$. An estimate of the standard deviation of the population is thus

$$2.6(0.325) = 0.84,$$

which agrees quite closely with the previous estimate of $s = 0.82$.

The "range times w" method always provides an unbiased estimate of σ. This means that, if the procedure is carried out on a number of different samples from the same population, the distribution of the estimates obtained has a mean value of σ. The usual estimate $s = \sqrt{\Sigma(y_i - \bar{y})^2/(n-1)}$, on the other hand, has a slight bias but a slightly higher efficiency.* In the circumstances presently discussed, either estimate has satisfactory properties but the estimate from the range has the advantage of being easy to calculate.

If the sample size is much larger than ten, the efficiency * of the range method falls off. A more efficient variation of the method can be employed for larger sample sizes. This involves dividing up the sample into equal subsamples of size 10 or less, but usually not less than four, by some random selection procedure, and using the average range of the subsamples to calculate an estimate of σ.

* The relative efficiency of one method of estimating compared with another method is obtained by comparing the numbers of observations needed by the two methods to produce estimates which have the same variance. The method which uses the fewer observations is the more efficient.

Example. Table 2.4 shows 32 observations from a certain process. They were allocated in a random manner to produce four groups each containing eight observations. The extreme values in each group are underlined and the ranges are given. The average of the four ranges is 65.75 and, using Table 2.3, we find that $w = 0.351$ for $n = 8$. The mean range estimate of σ is thus $65.75 \times 0.351 = 23.1$. The estimate obtained from the sum of squared deviations (Equation 2.5.2) is 25.7.

Table 2.4. Range estimate of σ for 32 observations

Observations	Range
65, 71, 43, 69, 52, *42*, *90*, 79	48
1, 57, 31, 44, 32, *70*, 63, 60	69
68, 50, 14, *70*, 55, 66, *6*, 31	64
14, 46, 22, 18, 85, 66, 88, *96*	82
Average range	65.75
Estimate of σ is $65.75 \times 0.351 = 23.1$	

A disadvantage of this mean range procedure is that the grouping is arbitrary, and different groupings of the observations lead to somewhat different estimates. In the use to which we later put this procedure we do not suffer this disadvantage, however, because the groups are completely specified by the EVOP scheme. (It should be noted that the procedure of the example is appropriate only when all the groups contain the same number of observations. For the procedure when the groups are unequal, see Section 5.5.)

2.6. DISTRIBUTION OF THE SAMPLE AVERAGE (OR SAMPLE MEAN)

When results subject to error are averaged, we expect the reliability of the average to increase as the number of observations on which the average is based increases. The exact manner in which this happens on the assumption that the observations are statistically independent can be studied by considering the *sampling distribution* of the sample average.

To understand what is meant by the sampling distribution of the sample average, once more we can think of the hypothetical population results as numbers written on tickets in a lottery drum. If the observations vary independently, then the taking of a sample of n observations is simulated by the random drawing of tickets from the drum. Suppose we are interested in the sampling distribution of the average of samples of size $n = 4$. We can imagine taking a sample of four tickets, calculating the average, replacing the sample, taking another sample of four tickets, calculating the average of this sample, and so on, until a very large number of samples of four have been taken. Now suppose that, as each sample average was obtained, its value was written on another ticket and that these "average tickets" were placed in a second drum. The tickets in the second drum would now represent the population of averages of four drawings from the first population. What would this second population be like? What kind of a distribution would it have? How would the characteristics of this sampling distribution depend on those of the original "parent" distribution? Suppose the mean of the original parent population was μ and its variance was σ^2. Then, whether the distribution of the observations in the parent population was normal or not:

1. The mean of the distribution of sample averages would be μ, the same as the mean of the parent population.
2. The variance of the distribution of sample averages would be σ^2/n, one nth the variance of the parent population. In symbols, if $V(y) = \sigma^2$, then $V(\bar{y}) = \sigma^2/n$, where V denotes variance, y denotes any single observation, and $\bar{y}$ denotes the average of n observations $y_1, y_2, \ldots, y_n$. It follows, of course, that the standard deviation of $\bar{y}$ (the square root of the variance) will be $\sigma/\sqrt{n}$.

If the distribution of the parent population were normal, so would be the distribution of the sample average.

Increase in Accuracy Produced by Averaging

We thus see precisely how the accuracy of the sample average regarded as an estimate of μ builds up as the number of independent observations in the sample is increased. Suppose, for example, that the individual yield for a production process varied randomly and normally about a mean of $\mu = 65$ with a standard deviation of $\sigma = 1$. Figure 2.17*a* shows the parent normal distribution for the individual results and the resulting normal distribution of averages of four (with standard deviation equal to one-half) and of averages of 16 (with standard deviation equal to one-fourth).

Tendency for the Distribution of an Average to Approach the Normal Form

In the foregoing illustration the parent distribution is supposed to be normal. Let us suppose now that the parent distribution is *nonnormal.*

Then the distribution of $\bar{y}$, although having a mean μ and standard deviation $\sigma/\sqrt{n}$, will not be exactly normal. However, the distribution will rapidly tend towards the normal form with increasing sample size. The reason is that

$$\bar{y} = \frac{y_1 + y_2 + \cdots + y_n}{n} = k_1 y_1 + k_2 y_2 + \cdots + k_n y_n,$$

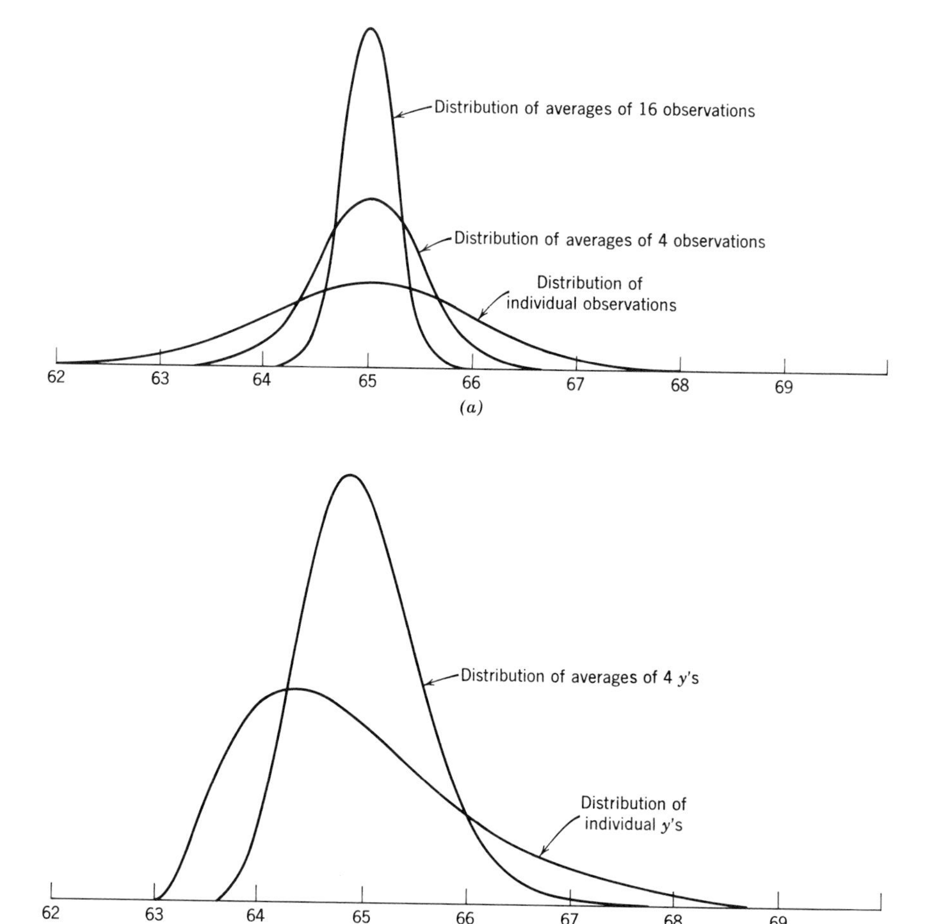

Figure 2.17. *(a) A normal distribution of individual observations from a production process with distributions for averages of 4 and of 16 observations. (b) Distributions of single observations and averages of four observations.*

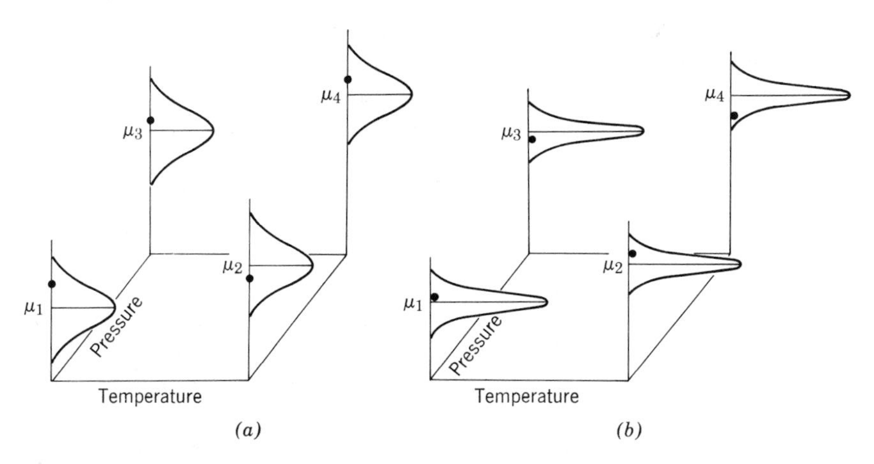

Figure 2.18. *Distributional changes as the number of cycles increases: (a) distributions of single observations (one cycle completed); (b) distributions of averages of four observations (four cycles completed).*

where all the k's equal $1/n$. The central limit theorem mentioned in Section 2.4 thus ensures that the distribution of $\bar{y}$ approaches the normal form as the sample size is increased. This is illustrated in Figure 2.17*b*, where we have deliberately chosen a parent distribution for the individual y's (with mean $\mu = 65$ and standard deviation $\sigma = 1$ as before) which is markedly nonnormal. It is seen to be heavily "skewed" to the right. The corresponding distribution of averages of four is distributed about the same mean with half the standard deviation of the parent distribution as expected. In addition it will be seen that the distribution of the sample means is much closer to the normal form. In particular it is much less skewed.

Happily, in our applications, we almost always wish to make probability statements about averages and not about individual results. Because of the central limit effect, even though the parent distribution is not exactly normal, the normal distribution tables will usually give good approximations for the distribution of averages.

Use of Averaging in Evolutionary Operation

In Figure 2.18 the four corners of the square base represent four particular combinations of temperature and pressure such as might be selected in an EVOP scheme. The vertical heights μ_1, μ_2, μ_3, μ_4 represent the true means for a response such as yield at the four sets of conditions. Figure 2.18*a* shows the distributions, about the true means, of single observations taken at each set of conditions. Also shown in the diagram are four dots representing actual values which might be obtained in a particular cycle. The spread of the distributions is such that, if only a single observation from each distribution were available, the random variation might obscure the real differences between the four sets of conditions. After four cycles of the phase, however (Figure 2.18*b*), an average of four results would be available at each set of conditions. The spreads of the error distributions of these averages would now be only one-half of those at the first cycle and differences between the four processes would be more apparent.

We see, then, that in order to distinguish real differences in the face of experimental error we need to consider the sample averages in relation to the spread of the distributions from which they come. When sufficient cycles have been performed to make the possible uncertainty in the sample averages reasonably small compared with differences between the averages, we are in a position to draw conclusions about the nature of the differences between the underlying true means.

2.7. MEAN VALUE AND VARIANCE OF IMPORTANT CONTRASTS

To monitor results from an EVOP project it is helpful to calculate, as a routine matter certain contrasts among the observations or averages of observations available. Thus for the EVOP scheme of Figures 2.18*a* and *b* the average yield results after *n* cycles might be as shown in Figure 2.18*c*. Among other things, we might wish to compare the over-all yield average of *high*-temperature runs with the over-all yield average of *low*-temperature runs. To do this we could calculate

$$L = \tfrac{1}{2}(68.0 + 68.2) - \tfrac{1}{2}(66.8 + 67.2) = 1.1.$$

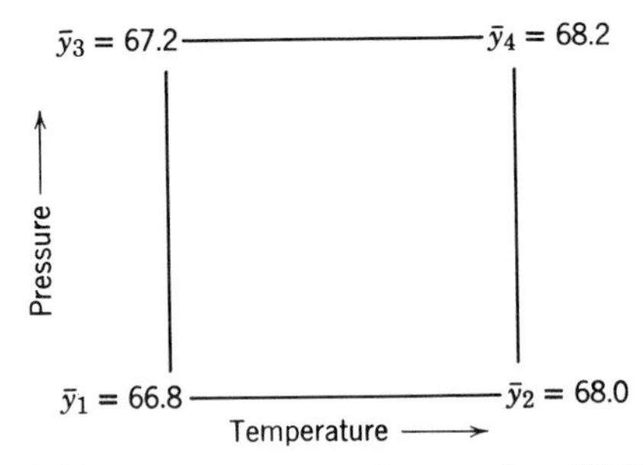

Figure 2.18c. Yield averages after n cycles of EVOP.

In this particular set of four cycles, then, high-temperature runs gave yields 1.1% higher on average than the low-temperature runs. As we shall see later, this contrast between high- and low-temperature runs is called the temperature *effect*. Now since the individual averages 68.0, 68.2, 66.8, and 67.2 are all subject to error, so too must be this calculated over-all temperature effect of 1.1.

In assessing the advisability of a permanent increase in temperature, we must be concerned with the question of how reproducible is the happening which has led to our estimate of a temperature effect of 1.1. Arguing as before, we can think of this happening as one realization from a hypothetical infinite population of such happenings. Central limit considerations will tell us that the distribution of the hypothetical population of L's can be expected to be approximately normal. We will know all that can be known about this distribution, therefore, if we know its mean value and variance. Certain simple rules exist by which we can find the mean and variance of quantities such as L, and these we now consider.

In symbols, the temperature effect can be written as

$$\begin{aligned} L &= \tfrac{1}{2}(\bar{y}_2 + \bar{y}_4) - \tfrac{1}{2}(\bar{y}_1 + \bar{y}_3) \\ &= -0.5\bar{y}_1 + 0.5\bar{y}_2 - 0.5\bar{y}_3 + 0.5\bar{y}_4, \end{aligned}$$

The quantities -0.5, $+0.5$, -0.5, $+0.5$ appearing in the expression are fixed constants. By contrast, the quantities $\bar{y}_1$, $\bar{y}_2$, $\bar{y}_3$, $\bar{y}_4$ are not fixed constants but vary from one realization to another. Quantities of this kind *which have a distribution* are called *random variables*. More generally, suppose there are p random variables $Y_1, Y_2, \ldots, Y_p$ and p constants $l_1, l_2, \ldots, l_p$. The expression

$$L = l_1Y_1 + l_2Y_2 + \cdots + l_pY_p \tag{2.7.1}$$

is called a *linear combination* of the random variables $y_1, y_2, \ldots, y_p$. In the foregoing example $p = 4$, the constants are $l_1 = -0.5$, $l_2 = 0.5$, $l_3 = -0.5$, $l_4 = 0.5$, and the random variables are $Y_1 = \bar{y}_1$, $Y_2 = \bar{y}_2$, $Y_3 = \bar{y}_3$, and $Y_4 = \bar{y}_4$.

Mean Value of L

The mean of L, denoted by μ_L, (as we might expect) is

$$\mu_L = l_1\mu_1 + l_2\mu_2 + \cdots + l_p\mu_p. \tag{2.7.2}$$

In our example, for instance, where $L = -\tfrac{1}{2}\bar{y}_1 + \tfrac{1}{2}\bar{y}_2 - \tfrac{1}{2}\bar{y}_3 + \tfrac{1}{2}\bar{y}_4$,

$$\mu_L = -\tfrac{1}{2}\mu_1 + \tfrac{1}{2}\mu_2 - \tfrac{1}{2}\mu_3 + \tfrac{1}{2}\mu_4,$$

where $\mu_1, \mu_2, \mu_3, \mu_4$ are the true mean responses at the four sets of conditions.

Before considering the variance of L which we denote by $V(L)$, we must remind the reader of the importance of the assumption of *statistical independence*. It will be recalled that the random variables $Y_1, Y_2, \ldots, Y_p$ would be said to be *statistically independent* or, equivalently, to be *independently distributed* if the probability that Y_i takes a value in some particular range does not depend on the values of the other Y's.

The expression for the mean value μ_L of the linear combination

$$L = l_1Y_1 + l_2Y_2 + \cdots + l_pY_p$$

is true irrespective of whether the Y's are independent or not.

Variance of *L*

In most of our EVOP applications we can assume that the Y_i are distributed independently. *If this is so*, then the variance of the linear combination $L = l_1Y_1 + l_2Y_2 + \cdots + l_pY_p$ is given by

$$V(L) = l_1^2\sigma_1^2 + l_2^2\sigma_2^2 + \cdots + l_p^2\sigma_p^2. \tag{2.7.3}$$

Example. A manufacturer uses two independent units to make a particular product. The daily production of the first is Y_1 tons/day and of the second Y_2 tons/day. Daily production rates vary randomly and independently of each other, and long-run averages and standard deviations are $\mu_1 = 60$ tons/day, $\sigma_1 = 3$ tons/day, $\mu_2 = 150$ tons/day, and $\sigma_2 = 4$ tons/day. It costs \$5/ton to manufacture with the first unit but only \$2/ton to manufacture with the second. What are the mean and variance of the daily production cost $L = 5Y_1 + 2Y_2$?

Here $\mu_1 = 60$, $\mu_2 = 150$, $\sigma_1 = 3$, $\sigma_2 = 4$, so that

$$\mu_L = 5 \times 60 + 2 \times 150 = 600,$$

$$V(L) = 25 \times 9 + 4 \times 16 = 289.$$

Thus the mean daily production cost is \$600 with standard deviation \$17.

In the special case in which all the Y's have the same variance σ_Y^2

$$V(L) = (l_1^2 + l_2^2 + \cdots + l_p^2)\sigma_Y^2. \tag{2.7.4}$$

Thus for the situation illustrated in Figure 2.18*c*, the variance of the temperature effect

$$L = -0.5\bar{y}_1 + 0.5\bar{y}_2 - 0.5\bar{y}_3 + 0.5\bar{y}_4$$

would be, after n cycles,

$$\begin{aligned} V(L) &= (0.25 + 0.25 + 0.25 + 0.25)\frac{\sigma^2}{n} \\ &= \frac{\sigma^2}{n}. \end{aligned}$$

Mean and Variance of a Sum and a Difference

We frequently need to know the mean and variance of the sum or the difference of two randomly varying quantities.

Let y and z be two independent random variables with means μ_y and μ_z and with variances σ_y^2 and σ_z^2. Then

$$\begin{array}{ll} \textit{Sum} & \textit{Difference} \\ \mu_{y+z} = \mu_y + \mu_z, & \mu_{y-z} = \mu_y - \mu_z, \\ \sigma_{y+z}^2 = \sigma_y^2 + \sigma_z^2. & \sigma_{y-z}^2 = \sigma_y^2 + \sigma_z^2. \end{array} \tag{2.7.5}$$

The formulas are both special cases of the general result of the last section; for example,

$$y - z = (1)y + (-1)z,$$

so that, using (2.7.2),

$$\mu_{y-z} = (1)\mu_y + (-1)\mu_z = \mu_y - \mu_z$$

$$V(y - z) = \sigma_{y-z}^2 = (1)^2\sigma_y^2 + (-1)^2\sigma_z^2 = \sigma_y^2 + \sigma_z^2.$$

It is at first surprising that the variance of the difference $z - y$ should be the same as that of the sum $z + y$. However, a little thought will show that this must be so because

$$V(-y) = V(y).$$

In all cases appropriate standard deviations are obtained by taking the positive square root of the corresponding variance.

Examples occur when it would be unrealistic to assume that the quantities Y_1, Y_2, $\ldots$, Y_p in the linear combination

$$L = l_1Y_1 + l_2Y_2 + \cdots + l_pY_p$$

were distributed independently. In this case the expression for the variance of L includes not only the variance of the individual Y's but also their *covariances*. Just as the variance of a random variable Y_i is measured by

$$V(Y_i) = \frac{\Sigma_l(Y_{il} - \mu_i)^2}{N},$$

so the covariance between Y_i and Y_j is measured by

$$\text{cov}\,(Y_i, Y_j) = \frac{\Sigma_l(Y_{il} - \mu_i)(Y_{jl} - \mu_j)}{N},$$

where Y_{il} ($l = 1, 2, \ldots$) denotes the lth member of the Y_i population. When Y_i and Y_j are independent, cov (Y_i, Y_j) will be zero. (Warning: the converse is not true.)

In general,

$$V(L) = l_1^2\sigma_1^2 + l_2^2\sigma_2^2 + \cdots + l_p^2\sigma_p^2 + 2l_1l_2 \operatorname{cov}(Y_1, Y_2) + 2l_1l_3 \operatorname{cov}(Y_1, Y_3) + \ldots$$
$$+ 2l_{p-1}l_p \operatorname{cov}(Y_{p-1}, Y_p).$$

In our EVOP application we shall be able to assume that all covariances are zero and thus the formula in which covariance terms are omitted can be used.

Standard Errors

We have seen how standard deviations may be calculated for "contrasts" or "effects" of interest which are linear combinations of the form

$$L = l_1Y_1 + l_2Y_2 + \cdots + l_pY_p.$$

In particular, if the variances of the Y's are all equal to σ_Y^2, then

$$\sigma_L = (l_1^2 + l_2^2 + \cdots + l_p^2)^{1/2}\sigma_Y.$$

If σ_Y is not precisely known, an estimate s_Y of σ_Y may be substituted, and we then refer to the estimated standard deviation of L,

$$\hat{\sigma}_L = (l_1^2 + l_2^2 + \cdots + l_p^2)^{1/2}s_Y,$$

as the *standard error* of L and write it as S.E.(L).

Calculation of Standard Errors

We shall need standard errors for *differences* between averages and for other functions. This presents no difficulty. To write down an expression for the standard error of a given quantity of interest, we proceed as follows:

1. Write down the standard deviation of the quantity of interest in terms of the standard deviation σ of individual observations [using (2.7.4)].
2. Substitute the estimate of σ in this expression.

For example, suppose we need the standard error of the difference between two sample averages $\bar{y}_2 - \bar{y}_1$, each based on n independent observations, and suppose we have an estimate s of σ. From (2.7.4) or, equivalently (2.7.5),

$$\sigma^2_{\bar{y}_2-\bar{y}_1} = V(\bar{y}_2 - \bar{y}_1) = \frac{\sigma^2}{n} + \frac{\sigma^2}{n} = \frac{2\sigma^2}{n},$$

$$\sigma_{\bar{y}_2-\bar{y}_1} = \frac{1.414\sigma}{\sqrt{n}}.$$

The standard error of $\bar{y}_2 - \bar{y}_1$ is therefore $1.414s/\sqrt{n}$, and the "2 S.E. limits" are

$$\bar{y}_2 - \bar{y}_1 \pm 2\,\frac{1.414s}{\sqrt{n}}.$$

2.8. MAKING INFERENCES IN THE PRESENCE OF UNCERTAINTY: SIGNIFICANCE TESTS AND CONFIDENCE INTERVALS

Significance Tests

Suppose we are examining a batch product which has been made under standard conditions for some considerable period of time. This extensive past experience has shown that individual observations of yield y vary approximately normally and independently about a mean $\mu_0 = 65$ with a standard deviation of $\sigma = 1$.

Suppose, now, that the plant is shut down for a period during which a modification is made which might be expected to improve the yield somewhat. After the modified process has been started up and some observations of yield have been made, a question might arise: Does the distribution of observations after modification have the same mean $\mu_0 = 65$ as it had before shutdown or has the distribution shifted so that the true mean is now equal to some other value μ_1 greater than $\mu_0 = 65$? We shall suppose that the standard deviation will be essentially the same after startup as before. A statistician might express the problem by saying that there were two hypotheses about the distribution of the observations of the modified process: *a null hypothesis* that this distribution was normal with mean $\mu = 65$ and standard deviation $\sigma = 1$, and an *alternative hypothesis* that the normal distribution had shifted location to some new mean value μ *greater* than $\mu_0 = 65$ but had the same standard deviation $\sigma = 1$. By a *null* hypothesis we mean a hypothesis of *no* effect; in this case, that the distribution was the same after startup as before.

Now suppose we had a sample of $n = 4$ observations made after startup which yielded a sample average of $\bar{y} = 65.4$. We could use these observations to *test* the null hypothesis that the mean μ of the distribution from whence they came was $\mu_0 = 65.0$ *against* the alternative that it had changed to some other value μ_1 greater than 65.0. To test the null hypothesis we tentatively suppose it true. If doing so makes the observed results reasonably plausible, there is no reason to doubt the null hypothesis and we say the result is *not statistically significant*. If, on the other hand, the adoption of the null hypothesis makes the occurrence of the observed data very implausible, then this suggests that we ought to abandon the null hypothesis in favor of an alternative which is plausible, and we say that the result *is* statistically significant.

Let us now apply this general line of argument to the present example. If the null hypothesis is tentatively regarded as true, then the four observations made after modification come from a normal distribution with mean

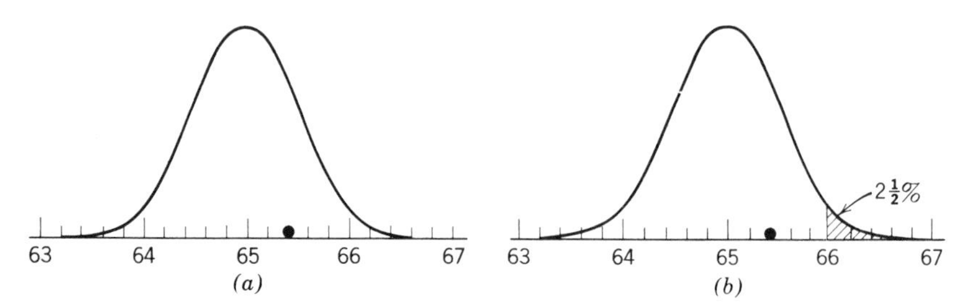

Figure 2.19. (*a*) *Reference distribution;* (*b*) *reference distribution with 2½% upper tail area shown.*

$\mu_0 = 65$ and standard deviation $\sigma = 1$. A sample average of $n = 4$ observations would therefore come from a normal* distribution with mean $\mu_0 = 65$ and standard deviation 0.5 (since $\sigma/\sqrt{n} = \sigma/\sqrt{4} = 0.5$). Figure 2.19*a* shows the observed sample average of 65.4 (indicated by the dot) in relation to a "reference" normal distribution centered at 65.0 with a standard deviation 0.5. Is it reasonable to regard $\bar{y} = 65.4$ as a typical member of this *reference distribution* appropriate under the null hypothesis? Since the average $\bar{y} = 65.4$ lies under the bulk of the reference distribution, our answer here must be "yes." The apparent divergence of $\bar{y} = 65.4$ from the mean $\mu_0 = 65.0$ is "not statistically significant." If the average $\bar{y}$ had been equal to, say, 67.0, our answer would have been "no," and we should have said that a significant discrepancy had occurred or, equivalently, that the null hypothesis had been discredited. We should be forced to the conclusion that the modification had induced a shift in the distribution in a favorable direction.

To make these rather general ideas quantitative, formal *levels* of significance may be introduced. This is usually done by considering the specific probability of getting a deviation of $\bar{y}$ from μ_0 in the anticipated direction as great or greater than that observed. From our knowledge of the normal distribution we know, for example, that positive "discrepancies" greater than 1.96 standard deviations occur 2.5% of the time. In Figure 2.19*b* the value $\bar{y} = 65.98$ [where $(65.98 - 65)/0.5 = 1.96$] just cuts off an upper tail area of 2.5% and thus corresponds to a value which is significantly high at the 2.5% level of significance. The value

$$\bar{y} = 66.29 \qquad \left[\text{where } \frac{(66.29 - 65)}{0.5} = 2.58\right]$$

* Because of the central-limit tendency of the sample means, the argument would be approximately correct even when the original distribution was moderately *nonnormal.*

would be significant at the 0.5% level. The 5% significance level is conventionally used to indicate a level of implausibility of the null hypothesis sufficient to cast doubt on its truth. The 1% significance level corresponds to an even greater degree of implausibility for the null hypothesis, and so on. The reader will appreciate that these levels are quite arbitrary and, at best, provide a rough guideline. If a significance test is to be used, it is best to state the actual probability that a deviation as great or greater than that observed will be found by chance. To assert only that "the result is significant at the 5% level" simply means that the tail probability is less than 5% and may be misleading.

Sequential Significance Tests

In the foregoing example we have supposed that we had a fixed number $n = 4$ of observations on which our analysis was to be based. In some circumstances observations might come to hand one by one, and we might decide *after each* observation whether we ought to (*a*) reserve judgment until at least one further observation was at hand, (*b*) decide that the mean μ after modification was unchanged and equal to μ_0 (not reject the null hypothesis), or (*c*) decide that the mean had in fact increased (reject the null hypothesis). A procedure of this kind is called a *sequential significance test.*

The sequential significance test takes account of the fact that observations are being repeatedly reviewed and, on the average, allows decisions to be reached with somewhat smaller sample sizes than is possible with fixed-sample-size tests.

Consider the example previously discussed of testing the null hypothesis that the true mean yield μ of the process had remained at its previous value $\mu = \mu_0 = 65$ against an alternative that μ had increased, i.e., $\mu > \mu_0$. We assume as before that the observations are approximately normally distributed with known standard deviation ($\sigma = 1$ in this case). To set up the sequential test we need to consider two additional quantities. These are (*a*) the *size* of difference δ, which it is "important" to detect, and (*b*) the probability β that we are prepared to tolerate of failing to detect this difference when it exists. Obviously these two quantities must be decided on together. In this example we might decide that we will be prepared to run the risk of $\beta = 0.025$ (2.5%) of failing to detect an improvement in yield of $\delta = 1.0$.

In summary then, we have the following table:

Symbol	General Meaning	For This Example
μ_0	The standard value	$\mu_0 = 65.0$
δ	The difference it is important to detect	$\delta = 1.0$
α	The probability of asserting a difference when none exists	$\alpha = 0.025$
β	The probability of asserting no difference when a difference δ exists	$\beta = 0.025$
σ	The standard deviation of the observations	$\sigma = 1$

To make the test we need the following quantities:

$$h_0 = \frac{-b\sigma^2}{\delta}, \qquad h_1 = \frac{a\sigma^2}{\delta}, \qquad s = \mu_0 + \tfrac{1}{2}\delta,$$

where

$$a = \ln \frac{1 - \beta}{\alpha}, \qquad b = \ln \frac{1 - \alpha}{\beta},$$

where ln refers to the natural logarithm. If, as is often convenient, we set $\alpha = \beta$, then $a = b$, and we find, for example,

$\alpha = \beta$:	0.05	0.025	0.01
$a = b$:	2.94	3.66	4.60

For the present example, then,

$$h_0 = -3.66, \qquad h_1 = 3.66, \qquad s = 65.50.$$

The procedure for carrying out the test may be stated as follows:

1. As each observation y comes to hand, calculate the quantity

$$r = y - s.$$

2. Keep a running total of the sum $S(r)$ of the r's.
3. So long as $S(r)$ lies between h_1 and h_0, continue taking observations. As soon as $S(r)$ exceeds h_1, terminate the test and conclude that a real improvement has occurred. As soon as $S(r)$ falls short of h_0, terminate the test and conclude that no improvement has occurred.

A possible sequence of five observations leading to the conclusion that a real improvement had occurred is shown in the following table.

y	$r = y - 65.5$	$S(r)$	Decision
66.7	1.2	1.2	$S(r)$ between h_0 and h_1, so continue testing
65.3	−0.2	1.0	$S(r)$ between h_0 and h_1, so continue testing
65.9	0.4	1.4	$S(r)$ between h_0 and h_1, so continue testing
67.2	1.7	3.1	$S(r)$ between h_0 and h_1, so continue testing
66.1	0.6	3.7	$S(r)$ greater than h_1, so conclude that a real improvement has occurred

Confidence Intervals

Significance tests are a somewhat overworked statistical tool. The formal mold of hypothesis testing sketched previously is not one that fits many

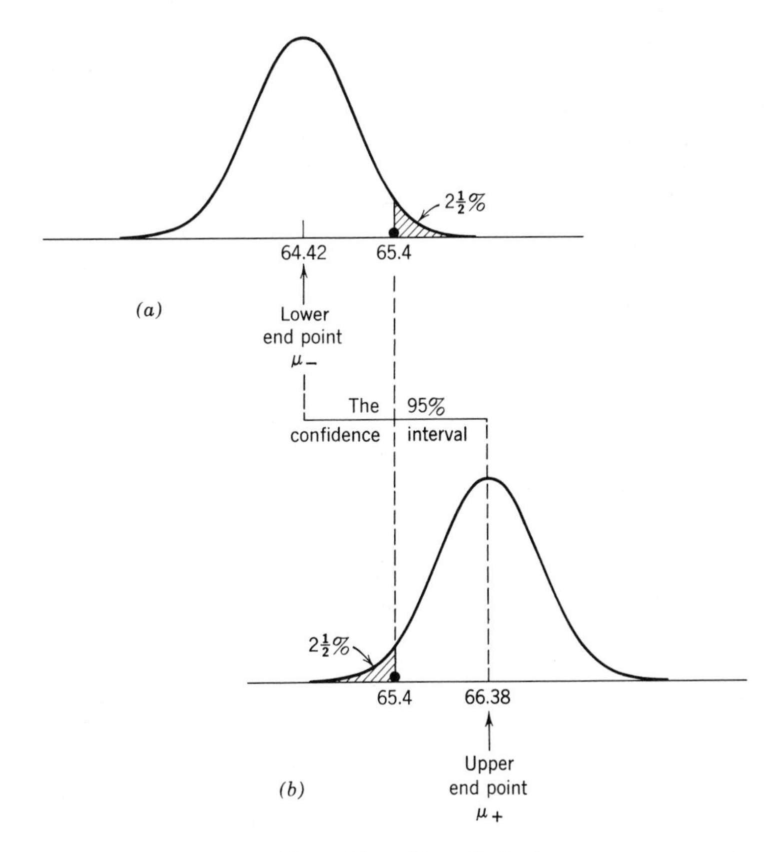

Figure 2.20. *95% confidence limits for μ.*

real situations. The investigator, aware as he is that his observed average $\bar{y}$ may be in error, is more likely to be interested in an *interval* about $\bar{y}$ which can be expected to include the true mean μ.

The problem of finding such an interval may be approached as follows. We have discussed previously the question of testing the hypothesis that the mean μ is equal to $\mu_0 = 65.0$ at a given level of significance, say, 2.5% against the *one-sided* alternative that $\mu > \mu_0$. We could equally well have tested the hypothesis that the true mean μ is equal to any other hypothesized value μ_0 against the *two-sided* alternative that $\mu > \mu_0$ or $\mu < \mu_0$.

Extending the idea illustrated in Figure 2.19, we can imagine the reference distribution slid backward and forward on the horizontal axis with the mean μ of the reference distribution taking various hypothetical values. There would be an interval of values of μ for which it would appear that $\bar{y}$ could plausibly have arisen from a reference distribution centered at μ. If we regard as plausible that range of values of μ which just fail to make the value $\bar{y}$ significant at the 2.5% level on each side, then this range is called a 95% confidence interval for μ and its end points are called the 95% *confidence limits* for μ.

From Figures 2.20*a* and *b* we see that the boundaries of the confidence interval would be the values μ_- and μ_+, for which

$$\frac{\bar{y} - \mu_-}{\sigma_{\bar{y}}} = +1.96 \qquad \text{and} \qquad \frac{\bar{y} - \mu_+}{\sigma_{\bar{y}}} = -1.96.$$

Thus μ_- and μ_+ are the values

$$\bar{y} \pm 1.96\sigma_{\bar{y}}.$$

For our example these boundaries are $65.4 \pm 1.96 \times 0.5$, so that the 95% confidence interval extends from 64.42 to 66.38. A 99% confidence interval would similarly be given by $65.4 \pm 2.58 \times 0.5$.

In general the $100(1 - \alpha)\%$ confidence interval for μ would be given by

$$\bar{y} \pm (z_{1-\alpha/2}) \frac{\sigma}{\sqrt{n}},$$

where $z_{(1-\alpha/2)}$ is the normal deviate which cuts off a right-hand tail area of $\frac{1}{2}\alpha$. It can be shown that such a confidence interval has the property that in repeated trials it will cover the true value μ a proportion $1 - \alpha$ of the time.

The *t* Distribution

The foregoing arguments have stemmed from the fact that the quantity $(\bar{y} - \mu)/(\sigma/\sqrt{n})$ follows a normal distribution with mean 0 and variance 1, (exactly if the original observations are normal, and approximately otherwise). When we do not know σ^2, we must use an estimate s^2 of it, instead. The distribution of the quantity $t = (\bar{y} - \mu)/(s/\sqrt{n})$ is rather like a normal distribution but has somewhat more probability in the tails. For example, the limits which enclose 95% of the *t*-distribution are somewhat further apart than the 95% limits of the corresponding normal distribution. This extra spread merely reflects the uncertainty with which s^2 estimates σ^2. The greater the number of observations on which s^2 is based, the less this uncertainty is and the closer the appropriate *t*-distribution approaches the normal distribution. Tables of the *t*-distribution can be used to make significance tests and to yield confidence intervals which take account of the fact that σ^2 is estimated. Sequential *t*-tests which take account of this uncertainty have also been developed [see O. L. Davies (1956)].

We mention the *t*-distribution here chiefly to explain, to those familiar with it, why we do *not* use it. We shall see that, for reasons we discuss later, we do not employ estimates of σ^2 based on a very small number of observations. In these circumstances it is sufficient for our purpose to substitute s for σ and use the normal distribution as an approximation.

Practical Assessment of Uncertainty in EVOP

It is quite apparent that some account must be taken of uncertainty in the running of any EVOP scheme. When there is an apparent average difference of, say, two units of yield in favor of a modified process *B* over a standard process *A*, the investigator needs to know if the evidence is such as would justify his making a permanent change to *B* or not. He may ask, "Is it 2 ± 0.2 or 2 ± 20?" How best can this evidence be presented to him? Since the investigator will reassess his evidence after each cycle of operation, it was originally thought that sequential tests would provide the most appropriate device. In practice, however, it was soon found that such test procedures were too clumsy for routine use under EVOP conditions. Usually the investigator will have to consider not one response such as yield but a whole variety of responses: yield, color, purity, etc. Keeping tally on a host of sequential tests is laborious, but, more important, to set up in advance differences δ "which it is important to detect" for each of these responses is difficult if not impossible. In practice such responses cannot be considered individually. A moderate reduction in purity might be tolerable if it were accompanied by an increase in yield or an improvement in color. It is quite impractical to attempt to foresee every contingency in advance.

Eventually it was found most practical merely to indicate 2 S.E. limits for each of the quantities of interest. As we have seen, the standard error (S.E.) is the estimated standard deviation of the quantity concerned. Attached to all quantities which appear on the EVOP board, then, are $\pm$ limits which are, in fact, $\pm$ twice the estimated standard deviation of the quantity concerned. These are interpreted informally as giving some general idea of how precisely the quantity is known.

If we were concerned with only a single quantity to be assessed after a fixed number of cycles and if σ were known, these limits would be, approximately, 95% confidence limits. In fact, we have to consider many quantities, we inspect results after each cycle, and σ is not precisely known but is estimated from data. The error limits so calculated do, however, provide a useful guide to the investigator as to what it is worthwhile "interpreting" and what he can dismiss as noise. It is these limits for the individual means that are displayed in Figure 1.6.

CHAPTER 3

The 2^2 and 2^3 Factorial Designs

In Chapter 1 we saw that information on how to improve an industrial process could be obtained by selecting a cycle of minor modifications, or variants, of the current process and repeatedly running these in sequence. In this chapter we consider the pattern of modifications to be used. This pattern is called the *design.*

Quantitative and Qualitative Variables

The process conditions which we may vary are defined by variables (or factors) of two kinds—quantitative and qualitative. *Quantitative variables* can be varied on a continuous scale. Examples are temperature, concentration, pressure, and agitation speed. *Qualitative variables* are concerned with attributes and cannot be varied on a continuous scale. Examples are type of catalyst, type of raw material, operator. Many of the variables studied in an EVOP program will be quantitative but we shall often want to test qualitative changes as well. For example, we may wish to check the effect of disconnecting a recycle loop the efficacy of which is doubted, or of switching between two different feeds.

3.1. FACTORIAL DESIGNS

Factorial designs are particularly useful arrangements which may be used to investigate either quantitative or qualitative variables. Suppose we were interested in investigating the effect of the quantitative variable, concentration; the qualitative variable, type of catalyst; and the quantitative variable, temperature. Suppose we decided to try three different concentrations—10%, 15%, 20%; catalyst manufactured by two companies—catalyst A, catalyst B; and two different temperatures—160°C, 170°C. Then a factorial arrangement would employ each of the 12 possible combinations of conditions:

Conditions	1	2	3	4	5	6	7	8	9	10	11	12
Concentration, %	10	20	30	10	20	30	10	20	30	10	20	30
Catalyst	*A*	*A*	*A*	*B*	*B*	*B*	*A*	*A*	*A*	*B*	*B*	*B*
Temperature, °C	160	160	160	160	160	160	170	170	170	170	170	170

In using this arrangement, we should say that we were studying the variable *concentration* at three levels, the variable *catalyst* at two levels* and the variable *temperature* at two levels. The arrangement would be called a $3 \times 2 \times 2$ factorial design.

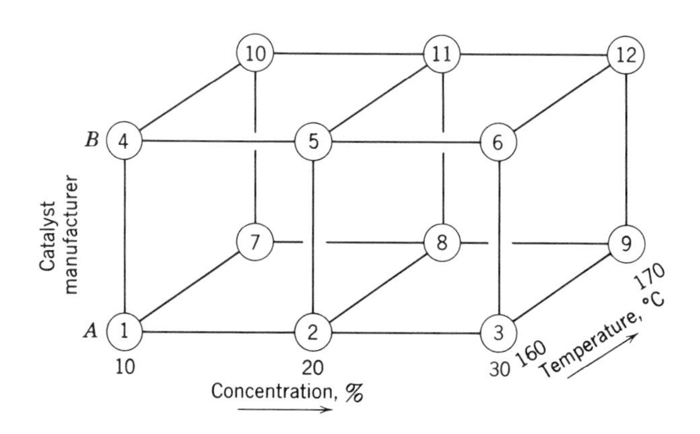

Figure 3.1. Representation of a $3 \times 2 \times 2$ factorial design.

Some of the advantages of such a design can be seen in Figure 3.1. For instance, a comparison of the results at conditions 7 and 1 would provide one measure of the effect of temperature at identical conditions of the other two variables (concentration 10%, catalyst *A*). The effect of temperature is equally measured by the comparison of results at conditions (8, 2), (9, 3), (10, 4), (11, 5), and (12, 6). This complete set of six comparisons allows us to determine the effect of temperature under all six tested combinations of concentration and catalyst. Thus all the results are being used to measure the effect of temperature.

It is equally true, however, that via the comparisons (4, 1), (5, 2), (6, 3),

* It has become standard parlance to use the word *level* to refer to the state of a qualitative as well as a quantitative variable.

(10, 7), (11, 8), and (12, 9) all results are being used to measure the effect of catalyst (again in each pair the level of the two remaining variables, concentration and temperature, are held constant).

Finally, via comparisons between (3, 2, 1), (6, 5, 4), (9, 8, 7), and (12, 11, 10) all the results may be used to measure the effect of changing the concentration.

The reader will see how the factorial design makes each result perform triple duty. Each piece of data supplies information on the effect of changing each one of the three variables. This feature is of particular value because it produces the maximum degree of averaging out of experimental error. Other features of factorial designs such as their ability to measure interaction will be appreciated as we proceed.

With EVOP a factorial design is run routinely under actual manufacturing conditions. This becomes progressively more difficult to do as the number of different conditions required by the design is increased.

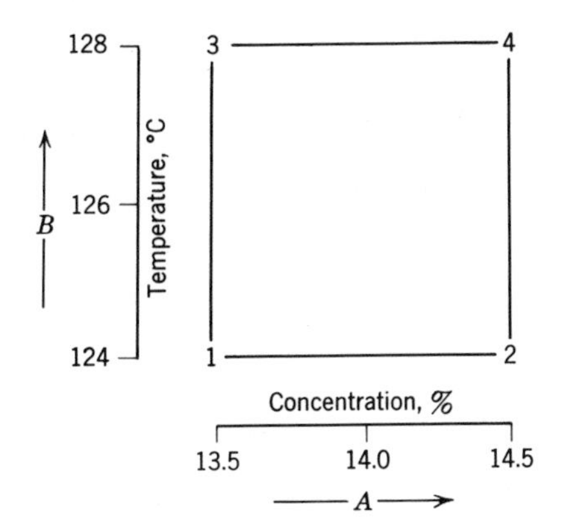

Figure 3.2. *A 2^2 factorial design in variables concentration and temperature.*

To ensure maximum simplicity EVOP has usually been run, therefore, with only two or three variables being simultaneously studied in any given phase and with each variable at only two levels. Thus the designs employed have been the 2^2 and 2^3 factorials, and we now make a detailed study of these extremely useful arrangements.

3.2. THE 2^2 FACTORIAL DESIGN

In the 2×2 or 2^2 factorial design we examine two variables A and B, each at two levels. Either of the variables may be quantitative or qualitative.

For illustration we return to an example introduced in Section 1.6 in which two quantitative variables, concentration (A) and temperature (B), were being studied. The four sets of process conditions forming the design are by convention shown at the vertices of a square as in Figure 3.2. The numbering system shown in the figure is convenient for discussing the design now, but a different numbering system more appropriate for sequencing the runs will be introduced later.

The averages for the response fluidity after four complete cycles were as follows:

Condition Number	Concentration, %	Temperature, °C	Fluidity (Averages after Four Cycles)
1	13.5	124	$\bar{y}_1 = 60.2$
2	14.5	124	$\bar{y}_2 = 67.6$
3	13.5	128	$\bar{y}_3 = 73.2$
4	14.5	128	$\bar{y}_4 = 76.2$

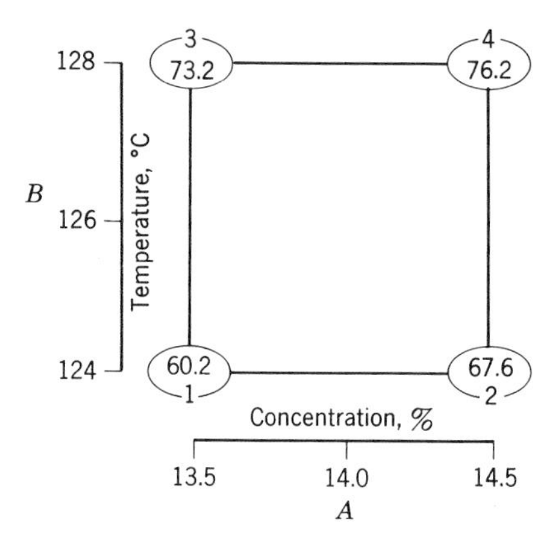

Figure 3.3. *Fluidity averages after four cycles of EVOP.*

The design can be represented as in Figure 3.2 as we have mentioned. By putting the fluidity averages actually on the diagram, we get Figure 3.3.

Effects for the 2^2 Factorial

Consider the difference $\bar{y}_4 - \bar{y}_3 = 76.2 - 73.2 = 3.0$. This number is an estimate of the effect on fluidity of changing concentration from 13.5 to 14.5% when the temperature is held at its high level of 128°. We can refer to this estimate, therefore, as the simple effect of concentration at the high level of temperature. Similarly, the simple effect of concentration when temperature is at its lower level is supplied by $\bar{y}_2 - \bar{y}_1 = 67.6 - 60.2 = 7.4$. The average of these two simple effects is called the *main effect* of concentration. It is usually indicated by a single capital letter symbol, in this case, A. Thus we have the following table:

Temperature Level, degrees	Concentration (A) Effect
128	$\bar{y}_4 - \bar{y}_3 = 3.0$
124	$\bar{y}_2 - \bar{y}_1 = 7.4$
Average = main effect $A = \frac{1}{2}(\bar{y}_4 - \bar{y}_3) + \frac{1}{2}(\bar{y}_2 - \bar{y}_1) = 5.2$	

The simple effects which are averaged are *contrasts* between results of high- and low-concentration runs. These contrasts are indicated in Figure 3.4*a* by arrows. Equivalently we see that the main effect of concentration is the average of the high-concentration results (indicated by plus signs in Figure 3.4*a*) less the average of the low-concentration results (indicated by minus signs in Figure 3.4*a*):

$$A = \tfrac{1}{2}(\bar{y}_4 + \bar{y}_2) - \tfrac{1}{2}(\bar{y}_3 + \bar{y}_1) = 71.9 - 66.7 = 5.2.$$

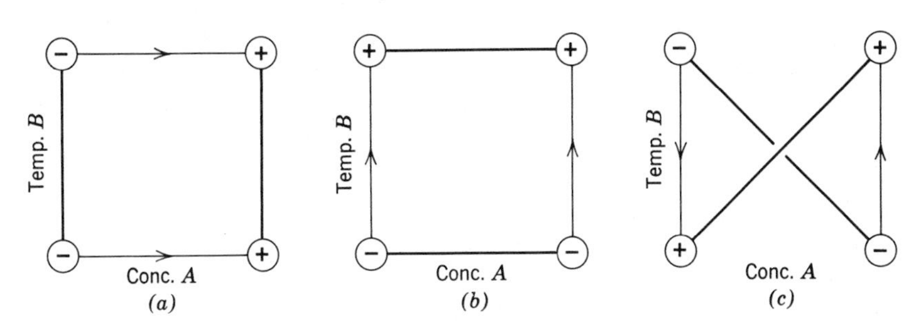

Figure 3.4. *Diagrammatic representation of effects for a two-level factorial:* (*a*) *concentration effect A;* (*b*) *temperature effect B;* (*c*) *interaction effect A × B.*

Similarly, for temperature, we have the following table:

Concentration Level	Temperature (B) Effect
14.5	$\bar{y}_4 - \bar{y}_2 = 8.6$
13.5	$\bar{y}_3 - \bar{y}_1 = 13.0$
Average = main effect $B = \frac{1}{2}(\bar{y}_4 - \bar{y}_2) + \frac{1}{2}(\bar{y}_3 - \bar{y}_1) = 10.8$	

The contrasts which are averaged are again shown by arrows in Figure 3.4*b*. Equivalently, as before, the main effect of temperature is defined as the average response at high temperature less the average response at low temperature,

$$B = \tfrac{1}{2}(\bar{y}_4 + \bar{y}_3) - \tfrac{1}{2}(\bar{y}_2 + \bar{y}_1)$$
$$= 74.7 - 63.9 = 10.8.$$

The high- and low-temperature runs which are contrasted are indicated by plus and minus signs in Figure 3.4*b*. We have so far derived from our data the following quantities:

Concentration main effect	$A = 5.2.$
Temperature main effect	$B = 10.8.$

Each of these effects utilizes *all* the data. The reader will notice that there are four repetitions at each of four sets of conditions, generating 16 observations in all, and that each of the main effects is the difference between two averages of eight results.

"Interaction" Between Variables

In the last table we have averaged two estimates of the temperature effect, 8.6 and 13.0, to arrive at the main effect of temperature. These two estimates are considerably different. Some difference would, of course, be expected because of experimental error. Even if there were no such error, however, we should expect that one variable would frequently have a different effect depending on the level of a second variable: for example, the effect produced by change in temperature would usually depend on the setting of other variables such as concentration and pressure. We discuss the effect of experimental error later. Ignoring it for the moment, let us consider how we might measure interaction with the data of the present example.

The contrasts $\bar{y}_4 - \bar{y}_2 = 8.6$ and $\bar{y}_3 - \bar{y}_1 = 13.0$ are the temperature effects at high and low concentration respectively. It is natural to measure the interaction in terms of their difference. By convention *half* this difference is defined as the interaction. Thus the measure of interaction AB between concentration A and temperature B is, for this example,

$$AB = \tfrac{1}{2}[(\bar{y}_4 - \bar{y}_2) - (\bar{y}_3 - \bar{y}_1)] = \tfrac{1}{2}(8.6 - 13.0) = -2.2.$$

We can once more think of this as an averaging of differences (indicated by the arrows in Figure 3.3*c*):

$$AB = \tfrac{1}{2}(\bar{y}_4 - \bar{y}_2) + \tfrac{1}{2}(\bar{y}_1 - \bar{y}_3) = \tfrac{1}{2}[8.6 + (-13.0)] = -2.2,$$

or a difference of averages indicated by plus and minus signs in Figure 3.3*c*:

$$AB = \tfrac{1}{2}(\bar{y}_4 + \bar{y}_1) - \tfrac{1}{2}(\bar{y}_3 + \bar{y}_2) = 68.2 - 70.4 = -2.2.$$

We see from the figure that we are, in fact, contrasting the averaged results on the "southwest-northeast" diagonal with the average results on the "southeast-northwest" diagonal.

General Interpretation of Main Effects in Relation to Interactions

We have now analyzed the 16 data values based on four repetitions of the four-point design in terms of main effects of factors A and B and an interaction AB as follows:

Main effect of concentration $A = \tfrac{1}{2}(\bar{y}_4 + \bar{y}_2 - \bar{y}_3 - \bar{y}_1) = 5.2.$
Main effect of temperature $B = \tfrac{1}{2}(\bar{y}_4 + \bar{y}_3 - \bar{y}_2 - \bar{y}_1) = 10.8.$
Interaction $AB = \tfrac{1}{2}(\bar{y}_4 + \bar{y}_1 - \bar{y}_3 - \bar{y}_2) = -2.2.$

In practice we would interpret apparent effects only if their standard errors indicated that such interpretation was justified. There would, for example, be little point in considering the possible meaning of the AB interaction of -2.2 if its two standard error limits (2 S.E. limits) were, say, ± 6. We take up this point later and suppose, for the moment, that the standard errors are small compared with the magnitude of the effects. The implication of the analysis is best understood by referring again to the two-way table of averages in Figure 3.3. Over the ranges studied, an increase in concentration and an increase in temperature separately produce increases in fluidity; their joint influence is, however, not as great as would be expected if the effects were additive.

We have seen that main effects are obtained by averaging simple effects. When these simple effects are, apart from experimental error, essentially the same, averaging results in reinforcement and the main effects thus indicate general trends and tendencies in the data. When these simple effects indicate markedly different phenomena (when there is powerful interaction), their averages considered *alone* have little meaning but must be interpreted in conjunction with the relevant interaction. Equivalently, we can say that, when substantial interaction occurs between factors, we must look at the joint *two-way table* of averages to interpret effects.

The following (rather extreme) example serves for illustration. Suppose, after a particular number of cycles, calculations produce the following results:

		Average Yield, %	
Modified catalyst		67	73
Normal catalyst		73	67
		Low	High
		Temperature	
2 S. E. Limits for averages		±1	
Effects with 2 S. E. Limits	Temperature effect	0 ± 1	
	Catalyst effect	0 ± 1	
	Temp. × catalyst interaction	6 ± 1	

In this example the main effect averages of the simple effects are zero, but this is because we are averaging large effects of opposite sign. As is indicated by the large interaction, these factors are nevertheless having a very powerful effect on yield. The nature of this effect is readily appreciated on inspection of the two-way table from which it is clear that, although a change in temperature from "low" to "high" level results in a drop in yield with a normal catalyst, it results in an increase with a modified catalyst.

A safe procedure for interpreting a factorial analysis is as follows: Suppose we have a main effect $A = 7 \pm 2$, where ± 2 refers to the 2 S.E. limits. We immediately inspect all two-factor interaction effects which involve A. (For a 2^2 factorial there would be only one such interaction AB but for a 2^3 factorial, for example, we would have to consider both AB and AC.) If all of the interaction effects involving A are negligible, then we can interpret the main effect directly. We can say that, for B, C, etc. within the ranges tested, the effect of changing A from a low level to a high level is 7 ± 2. If we had found a substantial interaction effect, say $AB = 4 \pm 2$, we should say that the effect of factor A depends on the level of B in the manner indicated in the appropriate two-way table of averages.

Periodic Reference to Current Best-known Conditions

As an EVOP program proceeds, there will exist, at any given stage, what may be called the *current best-known conditions*. At the beginning of the program these conditions will be those given in the operating specifications, but as improvements are incorporated the current best-known conditions will change. Potentially, each new phase can establish a new set of current best-known conditions. When a new phase is begun it is often helpful to be able to compare results directly with an appropriate reference. This can be done by returning to current best-known conditions once in every cycle.* In some instances this will mean that an additional point is added to the design, in others it will happen naturally that the current best-known conditions correspond to one of the standard design points. The idea is illustrated in Figure 3.5 again in relation to the quantitative variables, concentration and temperature. Figure 3.5*a* illustrates a case where current best-known conditions are included as part of the new design. This arrangement may be appropriate if there is information available suggesting that lower ranges of concentration and temperature should be investigated. Figure 3.5*b* illustrates a situation where current best-known conditions are outside the design. (This implies a less cautious move than that of Figure 3.4*a*. Without good reason a process superintendent might be justifiably reluctant to make it.) In Figure 3.5*c* current best-known conditions are at the center of the design. This is a particularly useful arrangement when no information is available as to the direction in which improvement should be sought.

The main advantage of inclusion of reference conditions is the reassurance this gives to the process superintendent. Anyone familiar with actual production conditions knows that the unexpected and unexplained frequently happens, even when no deliberate changes of any kind are made to the process. A quality characteristic may go almost out of specification

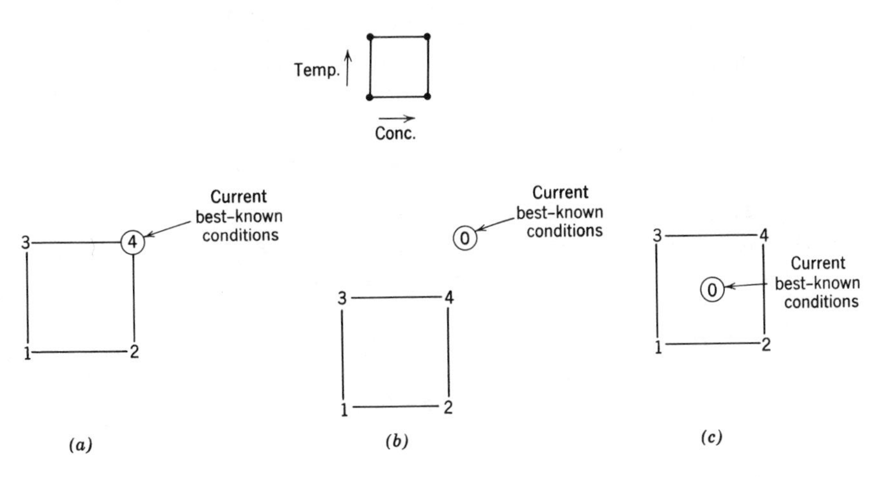

Figure 3.5. A 2^2 factorial design in relation to current best-known conditions

* When current best-known conditions use one of the sets of process conditions run in a previous phase, it might be thought unnecessary to rerun the conditions in the current phase. Rerunning is necessary because external influences such as quality of raw material and ambient temperature can affect the *general level* of a given response and so render inappropriate comparisons made with previous experience.

for a period and then mysteriously return to its normal value. After the initiation of EVOP there will be a great temptation to ascribe all such happenings to this program. The inclusion of reference conditions will help to clarify the situation.

When current best-known conditions are included in each cycle as a reference (whether as an additional point or as one of the factorial design points), it is informative to compare the average performance achieved while running the EVOP design with the performance which would have been obtained if all runs had been made at the reference conditions.

No matter where the reference point is located, such a comparison is always provided by the contrast:

Average response over all runs in the EVOP cycle

− average response at reference conditions.

This contrast is called the *change in mean* effect. Suppose n cycles have been completed so that for each set of conditions we can calculate an average response based on n results. Then with the numbering of the points shown in Figure 3.5, we have for arrangement (*a*),

$$\text{Change in mean} = \tfrac{1}{4}(\bar{y}_1 + \bar{y}_2 + \bar{y}_3 + \bar{y}_4) - \bar{y}_4$$
$$= \tfrac{1}{4}(\bar{y}_1 + \bar{y}_2 + \bar{y}_3 - 3\bar{y}_4).$$

For arrangements (*b*) and (*c*),

$$\text{Change in mean} = \tfrac{1}{5}(\bar{y}_1 + \bar{y}_2 + \bar{y}_3 + \bar{y}_4 + \bar{y}_0) - \bar{y}_0$$
$$= \tfrac{1}{5}(\bar{y}_1 + \bar{y}_2 + \bar{y}_3 + \bar{y}_4 - 4\bar{y}_0).$$

The change in mean provides a measure of the *direct* cost incurred by obtaining information in any particular phase. It is a measure of whether the conditions being run during the current phase yield an average result better or worse than the reference conditions. On the assumption that, had we not been running EVOP, we would have been operating statically at the current best-known conditions, it measures the temporary direct cost, if any, of running the scheme and so obtaining information during the present phase.

Phase Mean

An additional quantity which the process superintendent will find helpful is the *phase mean*. This is displayed on the information board just above the factor "effects." The phase mean is the mean response over all conditions being run in the present phase. It is estimated by the average of the results at all sets of conditions. Estimates of phase means for the various responses being investigated will indicate the average yield, quality, etc., of the material manufactured during a particular EVOP phase. Thus for the 2^2 design without additional reference conditions, the estimate of the phase mean is $\frac{1}{4}(\bar{y}_1 + \bar{y}_2 + \bar{y}_3 + \bar{y}_4)$ and, for the design with additional reference conditions, it is $\frac{1}{5}(\bar{y}_0 + \bar{y}_1 + \bar{y}_2 + \bar{y}_3 + \bar{y}_4)$. The *change in mean* is thus in all cases equal to *phase mean minus the mean at the reference conditions.*

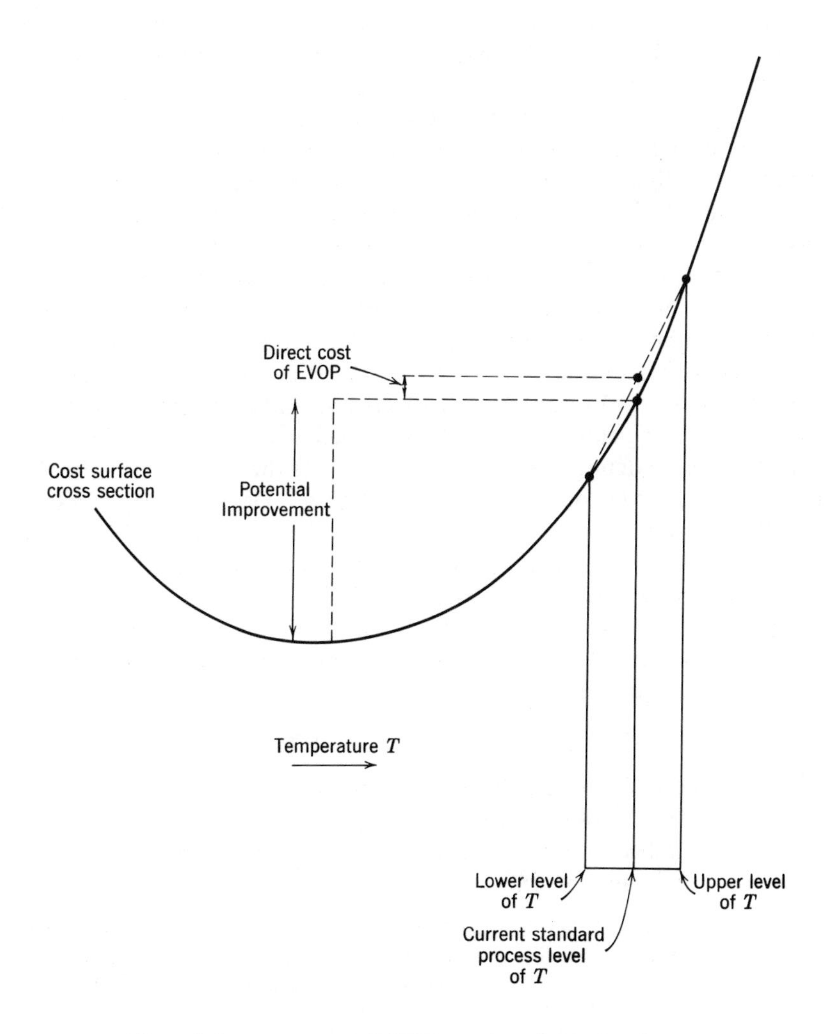

Figure 3.6. *Direct cost of EVOP; section of a cost surface.*

Change in Mean for a 2^2 Factorial Design With Center Point

In particular, when reference conditions are at the center of the factorial design, as shown in Figure 3.5*c*, a small direct cost is usually incurred. The reason for this can be seen by studying Figure 3.6. This shows a section of a possible cost surface for variation in the temperature T. Because of concavity of the cost surface, the cost at the center reference level of temperature is slightly lower than the average at the lower and upper levels of temperature. Thus running at the three levels of temperature costs slightly more than maintaining the center reference conditions. As in Figure 3.6, the potential *permanent* gain will, of course, often greatly exceed this *temporary* cost. Because of the *concavity* of the typical cost surface (Figure 3.6), the change in mean for cost would normally be positive and would estimate the slight local *increase* in cost. Similarly, because of the convexity of the typical yield surface, the change in mean yield would usually be *negative* and would estimate the slight local *loss* in yield.

In general, for a 2^2 factorial design with center point the change in mean is a measure of the surface curvature. Apart from the effect of experimental error, it will be positive near a minimum and negative near a maximum.* A change in mean effect which exceeds its 2 S.E. limits and is of about the same size numerically as the main effects may indicate that a maximum or minimum is being closely approached. Procedures for studying the response surface in more detail in such cases are outside the scope of this book but the interested reader should refer to Box (1954*b*), Davies (1956, chapter 11), and Cochran and Cox (1957, chapter 8*A*). In the common EVOP situations, often more can be gained by switching the study to other factors at this point than by detailed response studies of variables which are close to their optimal values. Again, using the fluidity data of Section 1.6 for illustration, the fluidity observed at the center conditions was 71.3, so that after four cycles the change in mean was

$$\tfrac{1}{5}[60.2 + 67.6 + 73.2 + 76.2 - 4(71.3)] = -1.6.$$

There is little to be learned from this particular example except, perhaps, that the surface is slightly convex. The large relative magnitudes of the main effects indicate, however, that if a maximum occurs for fluidity it is probably still fairly remote.

* For a 2^2 factorial design the true change in mean effect is actually a measure of the *sum* of the coefficients measuring quadratic curvature. We could usually assume that these coefficients were of the same sign (i.e., that we were approaching a "proper" minimum or maximum). A diagonally oriented saddle represented by equal coefficients of opposite sign could occur and yield a zero change in mean effect, but such a phenomenon would rarely be met in practice.

Sequencing of the Runs—Randomization

It would be normal statistical practice to randomize the order of the runs within each cycle. The effect of randomization is to make our procedures less dependent on assumptions. In particular, randomization ensures that, if systematic patterns or trends occur, arising from variables whose existence may not even be suspected, these will not be mistaken for effects of the deliberately introduced factors. It also would have the effect of validating our analysis (which assumes that errors within cycles are independent) in situations where the response errors are serially correlated. If runs can be randomized without too much difficulty, then this should be done. Our experience has been, however, that a random sequence is difficult to organize under actual production conditions and for the 2^2 design with added center points. We have usually employed the systematic sequence indicated by the numbering

$$\begin{array}{ccc} 4 & & 2 \\ & 0 & \\ 1 & & 3 \end{array}$$

In the actual conduct of such a scheme, it seems least confusing to number the conditions in the order in which they are run. It is this "sequence" numbering that we employ in subsequent discussion of this design and in making calculations on the model worksheets of Chapter 4.

Standard Errors for the Effects

We have shown how the data coming from the 2^2 factorial design with center point may be summarized in terms of main effects, an interaction and a change in mean effect. These quantities are, of course, subject to sampling error. Using the results of Section 2.7, we can easily obtain standard errors for these quantities. For example, the variance of the main effect of concentration (sequence numbering) is

$$\begin{aligned} V(A) &= V(\tfrac{1}{2}\bar{y}_2 + \tfrac{1}{2}\bar{y}_3 - \tfrac{1}{2}\bar{y}_4 - \tfrac{1}{2}\bar{y}_1) \\ &= \left(\frac{1}{4} + \frac{1}{4} + \frac{1}{4} + \frac{1}{4}\right)\frac{\sigma^2}{n} = \frac{\sigma^2}{n}, \end{aligned}$$

and, if we have an estimate s of σ, then

$$\text{S. E. }(A) = \frac{s}{\sqrt{n}}.$$

This is also the standard error for the main effect B and the interaction AB. The variance of the change in mean effect (sequence numbering) is

$$V(\tfrac{1}{5}\bar{y}_1 + \tfrac{1}{5}\bar{y}_2 + \tfrac{1}{5}\bar{y}_3 + \tfrac{1}{5}\bar{y}_4 - \tfrac{4}{5}\bar{y}_0)$$

$$= \left(\frac{1}{25} + \frac{1}{25} + \frac{1}{25} + \frac{1}{25} + \frac{16}{25}\right)\frac{\sigma^2}{n} = \frac{4}{5}\frac{\sigma^2}{n}.$$

Thus,

$$\text{S. E. (change in mean)} = \frac{2}{\sqrt{5}}\frac{s}{\sqrt{n}} = 0.89\frac{s}{\sqrt{n}}.$$

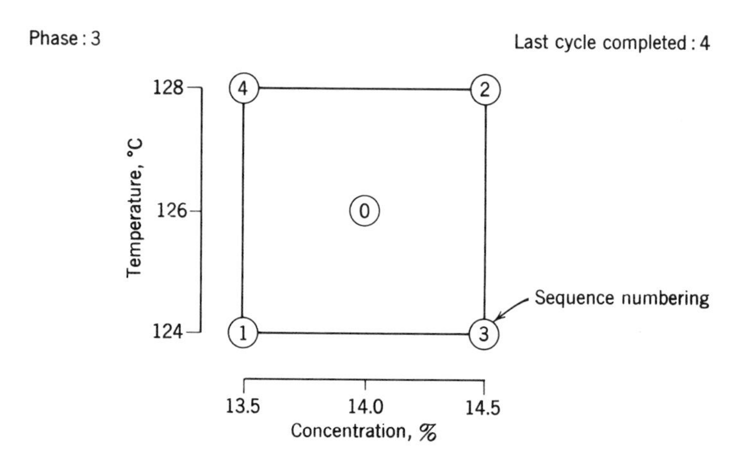

		Cost	Impurity, %	Fluidity
Requirement		Minimum	Less than 0.50	Between 55 & 70
Running averages		32.6 33.9 32.8 32.3 33.4	0.29 0.35 0.27 0.17 0.19	73.2 76.2 71.3 60.2 67.6
2 S. E. Error Limits		±0.7	±0.03	±1.1
Phase mean		33.0	0.25	69.7
Effects with 2 S. E. error limits	Concentration	1.2 ± 0.7	0.04 ± 0.03	5.2 ± 1.1
	Temperature	0.4 ± 0.7	0.14 ± 0.03	10.8 ± 1.1
	C × T	0.1 ± 0.7	0.02 ± 0.03	−2.2 ± 1.1
	Change in mean	0.2 ± 0.5	−0.02 ± 0.03	−1.6 ± 1.0
Standard deviations for individual observations		0.72	0.029	1.06

Figure 3.7. *Appearance of the information board at the end of four cycles.*

To complete the calculations, we need to obtain from the data an estimate s of the standard deviation σ. This is most easily done by a special procedure using the sample range. For convenience, the range procedure is not discussed until Section 3.3 after we have introduced the 2^3 design.

The Effects and Their Standard Errors for a 2^2 Design With Center Point

Finally, then, after n cycles, we could compute the following effects and standard errors:

	Effect (Sequence Numbering)	Standard Error
Main effect of concentration	$A = \frac{1}{2}(\bar{y}_2 + \bar{y}_3 - \bar{y}_4 - \bar{y}_1)$	$\frac{s}{\sqrt{n}}$
Main effect of temperature	$B = \frac{1}{2}(\bar{y}_2 + \bar{y}_4 - \bar{y}_3 - \bar{y}_1)$	$\frac{s}{\sqrt{n}}$
Interaction	$AB = \frac{1}{2}(\bar{y}_2 + \bar{y}_1 - \bar{y}_4 - \bar{y}_3)$	$\frac{s}{\sqrt{n}}$
Change in mean	$= \frac{\bar{y}_1 + \bar{y}_2 + \bar{y}_3 + \bar{y}_4 - 4\bar{y}_0}{5}$	$0.89 \frac{s}{\sqrt{n}}$

Analysis of Information Board for Three Responses Using the Factorial Effects

In our first introduction in Chapter 1 to the data of Figure 3.3, interpretation of the results for cost, impurity, and fluidity was based on an inspection of the individual running averages and their 2 S.E. limits. We have now seen how we can employ these averages to calculate the main effects, the interaction and the change in mean effect, and how we can obtain 2 S.E. limits for these quantities. We now illustrate how this additional analysis can be of practical value. Displayed in Figure 3.7 is the information board of Figure 1.6, but with additions showing the estimated effects of concentration and temperature, the concentration-temperature interaction, and the change in mean effect, as well as their 2 S.E. limits for each of the three responses: cost, impurity, and fluidity. The phase mean estimates are shown in Figure 3.7 without 2 S.E. limits. If limits are required for these quantities, they should be based on an estimate of the standard deviation *between cycles* and not on s which is a *within cycles* estimate. A method for obtaining an estimate of the standard deviation between cycles is given in Section 5.5.

For the responses *cost* and *impurity* the interactions AB are small in magnitude, both compared with their standard errors and also in relation to the magnitude of the main effects for A and B. Also, for both these responses, the changes in mean effects are small, i.e., the direct costs of information gathering or the over-all curvature of the surfaces are undetectable. For these responses, the changes that occur can be explained locally in terms of simple linear trends. To indicate trends the reality of which has been established, it is sometimes helpful to rough in contours as in Figure 3.8. The contours have been drawn by rough eye interpolation among the design points. It should be emphasized that no attempt need be made here to fit a response function formally. It is merely suggested that, having established by our analysis that we have effects to explain, eye-drawn contours can assist the investigator in deciding what ought to be done next. Where main effects dominate, as for cost and impurity, it is implied that the response function is locally planar and its contours are, therefore, equally spaced straight lines.

An appreciable AB interaction does occur for the response *fluidity*. As we have noticed already, this interaction is not, however, of the extreme kind where simple effects of the factors are of opposite sign, but merely indicates some lack of additivity in the effects. It is to be noted also that a negative *change in mean* effect is obtained of -1.6 ± 1.0. The suggestion is, therefore, that we are here dealing with a somewhat curved surface possibly approaching a maximum. Rough interpolation among the points yields the approximate 75, 70, and 65 contours shown in the figure. Since fluidity must be above 55, special interest attaches to the 55 contours for fluidity. This is outside the ranges of the design, but *to aid our thoughts*

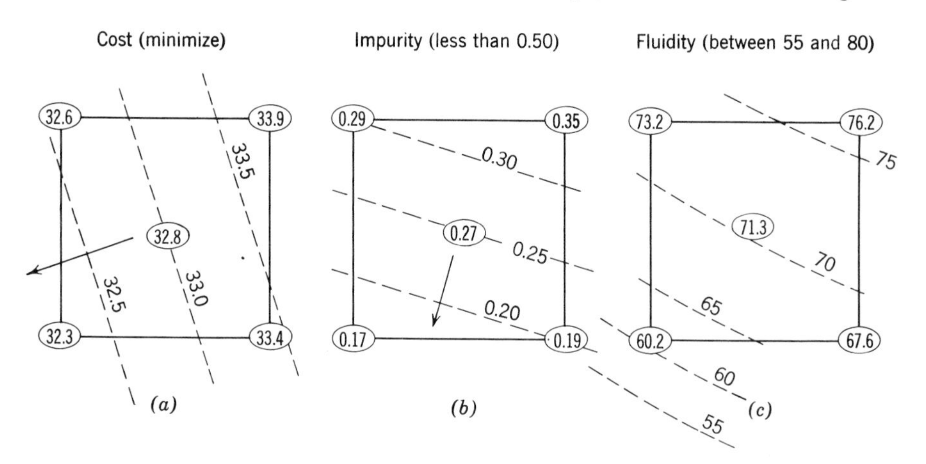

Figure 3.8. *Approximate "eye-drawn" contours for cost, impurity, and fluidity.*

on what ought to be done next (but certainly not for drawing definite *conclusions*), possible positions of the 60 and 55 contours are shown on the diagram.

Inspection of the diagram shows that, on our present guess as to the position and direction of the 55 contours for fluidity, we will do well to investigate the effect of reducing concentration, maintaining temperature at its median level of 126°C. As we explained in our earlier analysis of this example, the three sets of conditions (13%, 126°C), (13.5%, 126°C) and (14%, 126°C) were run in succession, resulting in the conditions (13%, 126°C) being used as the starting point for further investigation.

3.3. THE 2^3 FACTORIAL DESIGN

As we have explained, the difficulty of operating an EVOP scheme increases greatly as the number of factors is increased, because of the number of different process conditions involved and the number of changes that must be made. In actual experience it has been found, however, that a scheme which simultaneously examines three variables is perfectly feasible; in fact, some of the most successful EVOP schemes have involved three factors. To examine three factors we make use of the powerful 2^3 factorial design, which is now discussed in some detail.

As before, the factors involved may be quantitative, such as temperature, concentration, pressure, or qualitative, such as type of catalyst or type of raw material. For illustration we consider a 2^3 factorial design used to study the effects of temperature A, concentration B, and pressure C on the response: yield of by-product. The variables were run at the following levels:

	Low	High
Temperature A, °C	170	180
Concentration B, %	22	28
Pressure C, psi	30	40

The resulting 2^3 factorial design may then be represented by the eight vertices of a cube, numbered 1 to 8,* as shown in Figure 3.9. The eight sets of conditions, which include all combinations of the pairs of levels, are shown in Table 3.1. Also shown in this table are the average yields of by-product

* As for the 2^2 factorial, the numbering scheme used in the discussion of the properties of this design, and usually referred to as *standard order*, allows the effects and interactions to be defined with least notational confusion. For the model worksheet calculation described in Chapter 5, however, a different numbering which refers to the *sequence order* is used.

Table 3.1. The eight sets of conditions of a 2^3 factorial design

Standard Order	Temperature A	Concentration B	Pressure C	Average By-product Yields (Three Cycles)
1	170	22	30	3.9
2	180	22	30	4.8
3	170	28	30	2.9
4	180	28	30	3.6
5	170	22	40	3.5
6	180	22	40	5.1
7	170	28	40	2.8
8	180	28	40	4.4

after three cycles, rounded to a single decimal. In Figure 3.10, the average by-product yields are shown next to the appropriate corner of the cube.

Estimates of Main Effects

Now suppose we wish to obtain the effect of changing temperature A from the low to the high level under fixed conditions of the other variables. One such measure is provided by the difference between the yields from the second and first runs, $\bar{y}_2 - \bar{y}_1 = 4.8 - 3.9 = 0.9$. It will be noticed that in runs 1 and 2 both concentration and pressure are held fixed at their

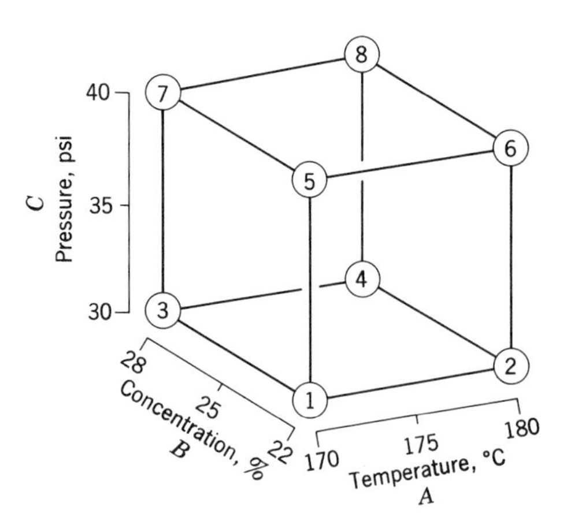

Figure 3.9. A 2^3 factorial design.

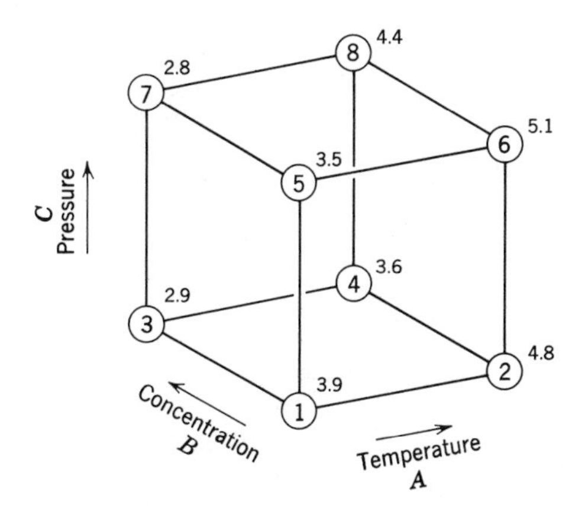

Figure 3.10. *A 2^3 factorial design with average by-product yields after three cycles.*

lower levels, and these conditions differ only in *temperature* level. Two more runs in which only temperature is changed are runs 4 and 3, yielding a second estimate of the temperature effect, $\bar{y}_4 - \bar{y}_3 = 3.6 - 2.9 = 0.7$. In all there are four such pairs and these are set out in Table 3.2.

As in the two-variable case, the average of these simple effects is called the *main effect*. Thus

main effect of temperature

$$= A = \tfrac{1}{4}[(\bar{y}_2 - \bar{y}_1) + (\bar{y}_4 - \bar{y}_3) + (\bar{y}_6 - \bar{y}_5) + (\bar{y}_8 - \bar{y}_7)] = 1.20.$$

Alternatively we can write,

$$\begin{aligned} A &= \tfrac{1}{4}(\bar{y}_2 + \bar{y}_4 + \bar{y}_6 + \bar{y}_8) - \tfrac{1}{4}(\bar{y}_1 + \bar{y}_3 + \bar{y}_5 + \bar{y}_7) \\ &= \tfrac{1}{4}(4.8 + 3.6 + 5.1 + 4.4) - \tfrac{1}{4}(3.9 + 2.9 + 3.5 + 2.8) \\ &= 4.475 - 3.275 = 1.20. \end{aligned}$$

Table 3.2. The main effect A of temperature

Measure of Temperature Effect A		Value	Level of Concentration B	Level of Pressure C
$\bar{y}_2 - \bar{y}_1$	$4.8 - 3.9$	0.9	Low	Low
$\bar{y}_4 - \bar{y}_3$	$3.6 - 2.9$	0.7	High	Low
$\bar{y}_6 - \bar{y}_5$	$5.1 - 3.5$	1.6	Low	High
$\bar{y}_8 - \bar{y}_7$	$4.4 - 2.8$	1.6	High	High
Average = temp. main effect A =		1.20		

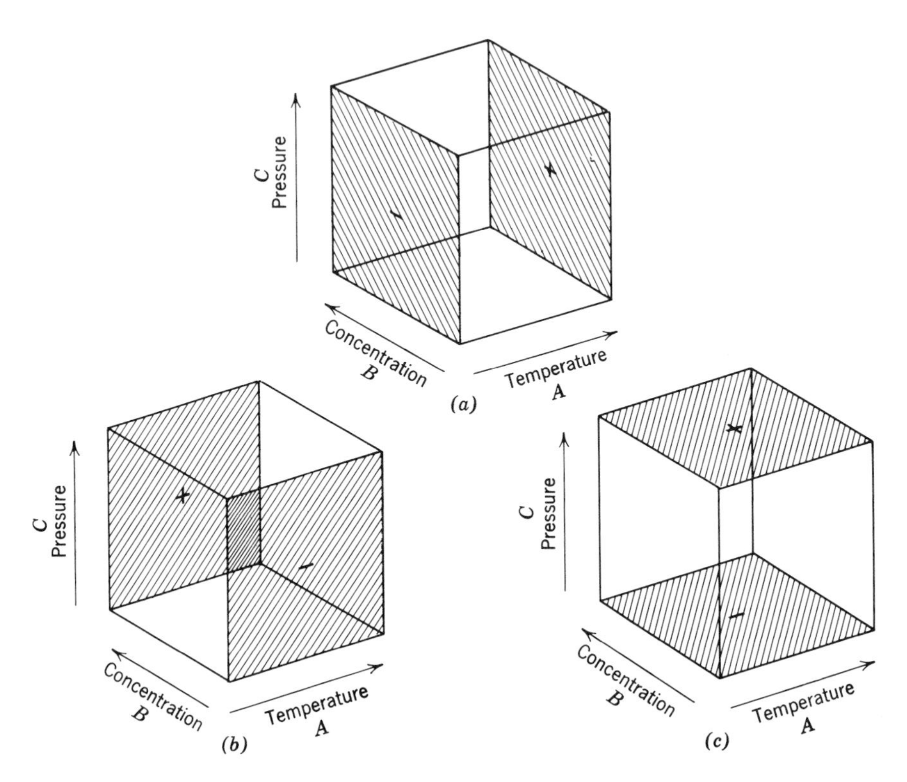

Figure 3.11. *Diagrammatic representation of the main effect contrast:* (*a*) *A effect;* (*b*) *B effect;* (*c*) *C effect.*

Thus the main effect is a comparison of the average response on the "high-temperature" face of the cube with the average on the "low-temperature" face (see Figure 3.11*a*). Similarly we see that, by comparison of the yields for runs 3 and 1, 4 and 2, 7 and 5, and 8 and 6, we obtain four separate measures of the effect of concentration *B*. Each of these comparisons are again made under identical conditions of the other two factors *A* and *C*. Specifically,

$$B = \tfrac{1}{4}[(\bar{y}_3 - \bar{y}_1) + (\bar{y}_4 - \bar{y}_2) + (\bar{y}_7 - \bar{y}_5) + (\bar{y}_8 - \bar{y}_6)].$$

Main effect of concentration *B*

$$= \tfrac{1}{4}[(2.9 - 3.9) + (3.6 - 4.8) + (2.8 - 3.5) + (4.4 - 5.1)]$$

$$= -0.90.$$

Again this is equivalent (Figure 3.11*b*) to comparing the average response on the high-concentration side of the cube with that on the low-concentration side for, after rearrangement of the foregoing expression, we obtain

main effect of concentration B

$$\begin{aligned} &= \tfrac{1}{4}(\bar{y}_3 + \bar{y}_4 + \bar{y}_7 + \bar{y}_8) - \tfrac{1}{4}(\bar{y}_1 + \bar{y}_2 + \bar{y}_5 + \bar{y}_6) \\ &= \tfrac{1}{4}(2.9 + 3.6 + 2.8 + 4.4) - \tfrac{1}{4}(3.9 + 4.8 + 3.5 + 5.1) \\ &= 3.425 - 4.325 \\ &= -0.90 \quad \text{as before.} \end{aligned}$$

In the same way

main effect of pressure C

$$\begin{aligned} &= \tfrac{1}{4}[(3.5 - 3.9) + (5.1 - 4.8) + (2.8 - 2.9) + (4.4 - 3.6)] \\ &= 0.15. \end{aligned}$$

This effect (Figure 3.11*c*) is the contrast between averages of observations on the high- and low-pressure faces of the cube. Specifically, on rearrangement, we have

main effect of pressure C

$$\begin{aligned} &= \tfrac{1}{4}(\bar{y}_5 + \bar{y}_6 + \bar{y}_7 + \bar{y}_8) - \tfrac{1}{4}(\bar{y}_1 + \bar{y}_2 + \bar{y}_3 + \bar{y}_4) \\ &= \tfrac{1}{4}(3.5 + 5.1 + 2.8 + 4.4) - \tfrac{1}{4}(3.9 + 4.8 + 2.9 + 3.6) \\ &= 3.95 - 3.80 \\ &= 0.15. \end{aligned}$$

Advantages Over "One-Factor-at-a-Time" Design

Some of the advantages of the factorial design over the "one-factor-at-a-time" design will now be clear. The one-factor-at-a-time procedure would have consisted of holding variables B and C fixed at some arbitrary chosen level and making runs at the upper and lower levels of A. Then A and C would be held fixed and runs would be made at the upper and lower levels of B. Finally, A and B would be held fixed and runs made at the upper and lower levels of C. Now such a procedure would only have been appropriate if the factors A, B, and C acted independently on the response (i.e., there were no interactions). But if this were so, then, by using the factorial, we would increase the efficiency of our experimentation threefold. This is so because, when we use the factorial design, we have the precision obtainable from the contrast between one average of four observations and another such average for *each* of the factors A, B, and C. To have obtained the same precision from a one-factor-at-a-time experiment would have required not 8 runs but 24 runs. The advantage of the factorial design over the one-factor-at-a-time design is, however, greater than is implied by the foregoing discussion. When the factors "interact," the one-factor-at-a-time design may be quite misleading, but the factorial arrangement will allow the interactions to be estimated and correct conclusions to be drawn.

The Two-factor Interactions *AB*, *AC*, and *BC*

The interaction between any two factors such as temperature A and pressure C is given by

$$AC = \tfrac{1}{2}[(\text{average effect of factor } A \text{ at the high level of } C) - (\text{average effect of factor } A \text{ at the low level of } C)].$$

Consider again the results in Table 3.2 which may be summarized in the following way:

Temperature A Effect			Level of Concentration B	Level of Pressure C
$\bar{y}_2 - \bar{y}_1$	0.9	0.8	Low	Low
$\bar{y}_4 - \bar{y}_3$	0.7		High	Low
$\bar{y}_6 - \bar{y}_5$	1.6	1.6	Low	High
$\bar{y}_8 - \bar{y}_7$	1.6		High	High

Averaging over concentration (Table 3.3), we obtain 1.6 as the average effect of temperature at high pressure and 0.8 as the average effect of temperature at low pressure. The difference $1.6 - 0.8 = 0.8$ is obviously a measure of the extent to which the temperature effect is different at the different pressures investigated, i.e., of the *interaction* between temperature and pressure. As for the 2^2 design, the interaction is defined as

Table 3.3. Calculation of the temperature-pressure interaction AC

Average Temperature (A) Effect	Pressure (C)
$\tfrac{1}{2}[(\bar{y}_2 - \bar{y}_1) + (\bar{y}_4 - \bar{y}_3)] = 0.8$	Low
$\tfrac{1}{2}[(\bar{y}_6 - \bar{y}_5) + (\bar{y}_8 - \bar{y}_7)] = 1.6$	High
Interaction $AC = \tfrac{1}{2}(1.6 - 0.8) = 0.4$	

half this difference and is called the *two-factor* interaction AC between the factors temperature A and pressure C. It can equally well be defined as

$$AC = \tfrac{1}{2}[(\text{average effect of factor } C \text{ at the high level of } A) - (\text{average effect of factor } C \text{ at the low level of } A)].$$

The reader can easily verify that this second definition gives precisely the same result as the first.

If we consider how we have manipulated the various quantities (Table 3.3) to arrive at the estimate of AC, we find that

$$\begin{aligned} AC &= \tfrac{1}{4}[(\bar{y}_6 - \bar{y}_5) + (\bar{y}_8 - \bar{y}_7) - (\bar{y}_2 - \bar{y}_1) - (\bar{y}_4 - \bar{y}_3)] \\ &= \tfrac{1}{4}[(5.1 - 3.5) + (4.4 - 2.8) - (4.8 - 3.9) - (3.6 - 2.9)] \\ &= 0.4. \end{aligned}$$

Again this may be written as a contrast between two means of four results:

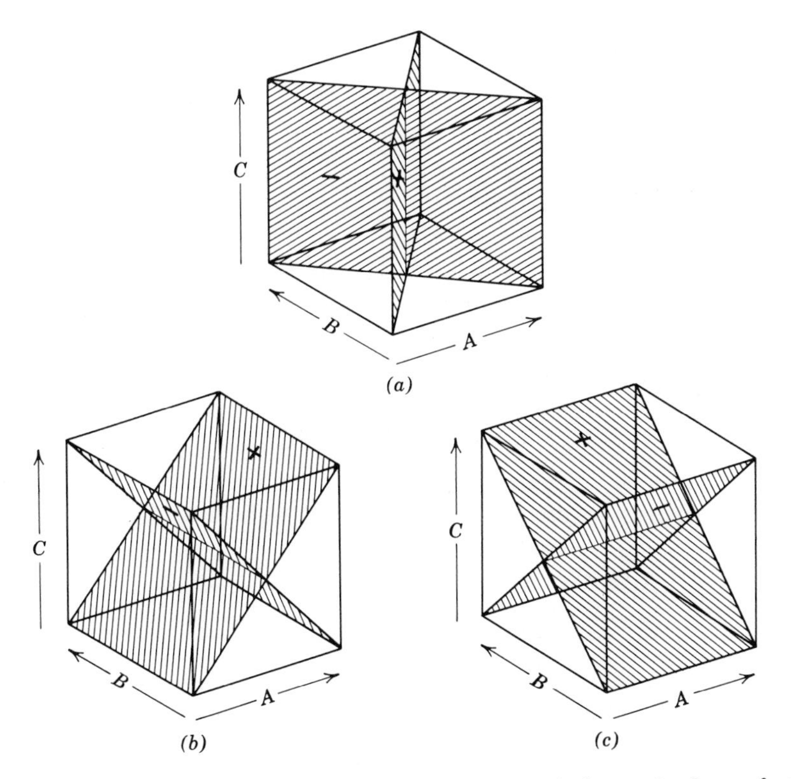

Figure 3.12. *Diagrammatic representation of the comparisons which comprise the two-factor interactions: (a) interaction AB; (b) interaction AC; (c) interaction BC.*

$$
\begin{aligned}
AC &= \tfrac{1}{4}(\bar{y}_1 + \bar{y}_3 + \bar{y}_6 + \bar{y}_8) - \tfrac{1}{4}(\bar{y}_2 + \bar{y}_4 + \bar{y}_5 + \bar{y}_7) \\
&= \tfrac{1}{4}(3.9 + 2.9 + 5.1 + 4.4) - \tfrac{1}{4}(4.8 + 3.6 + 3.5 + 2.8) \\
&= 4.075 - 3.675 \\
&= 0.40.
\end{aligned}
$$

From Figure 3.12*b* it is seen that the interaction AC is, in fact, a contrast between the averages of two sets of four results lying on two oblique planes which intersect on the cube's AC faces. Exactly similar arguments are used to define the AB and BC interactions, which turn out (Figures 3.12*a* and 3.12*c*) to be contrasts of means of four results on appropriate diagonal planes of the cube.

The three two-factor interactions are therefore

$$
\begin{aligned}
AB &= \tfrac{1}{4}(\bar{y}_1 + \bar{y}_4 + \bar{y}_5 + \bar{y}_8) - \tfrac{1}{4}(\bar{y}_2 + \bar{y}_3 + \bar{y}_6 + \bar{y}_7) \\
&= \tfrac{1}{4}(3.9 + 3.6 + 3.5 + 4.4) - \tfrac{1}{4}(4.8 + 2.9 + 5.1 + 2.8) \\
&= 3.85 - 3.90 \\
&= -0.05. \\
AC &= \tfrac{1}{4}(\bar{y}_1 + \bar{y}_3 + \bar{y}_6 + \bar{y}_8) - \tfrac{1}{4}(\bar{y}_2 + \bar{y}_4 + \bar{y}_5 + \bar{y}_7) \\
&= \tfrac{1}{4}(3.9 + 2.9 + 5.1 + 4.4) - \tfrac{1}{4}(4.8 + 3.6 + 3.5 + 2.8) \\
&= 4.075 - 3.675 \\
&= 0.40. \\
BC &= \tfrac{1}{4}(\bar{y}_1 + \bar{y}_2 + \bar{y}_7 + \bar{y}_8) - \tfrac{1}{4}(\bar{y}_3 + \bar{y}_4 + \bar{y}_5 + \bar{y}_6) \\
&= \tfrac{1}{4}(3.9 + 4.8 + 2.8 + 4.4) - \tfrac{1}{4}(2.9 + 3.6 + 3.5 + 5.1) \\
&= 3.975 - 3.775 \\
&= 0.20.
\end{aligned}
$$

An Estimate of the Three-factor Interaction *ABC*

For a 2^3 factorial design set up to investigate the factors A, B, and C, we have seen how to use the averages at the eight sets of conditions to

Table 3.4. Calculation of the ABC interaction

AB Interaction	Level of *C*
$\tfrac{1}{2}(\bar{y}_1 + \bar{y}_4) - \tfrac{1}{2}(\bar{y}_2 + \bar{y}_3) = -0.1$	Low
$\tfrac{1}{2}(\bar{y}_5 + \bar{y}_8) - \tfrac{1}{2}(\bar{y}_6 + \bar{y}_7) = 0.0$	High
Interaction $ABC = \tfrac{1}{2}[0.0 - (-0.1)] = 0.05$	

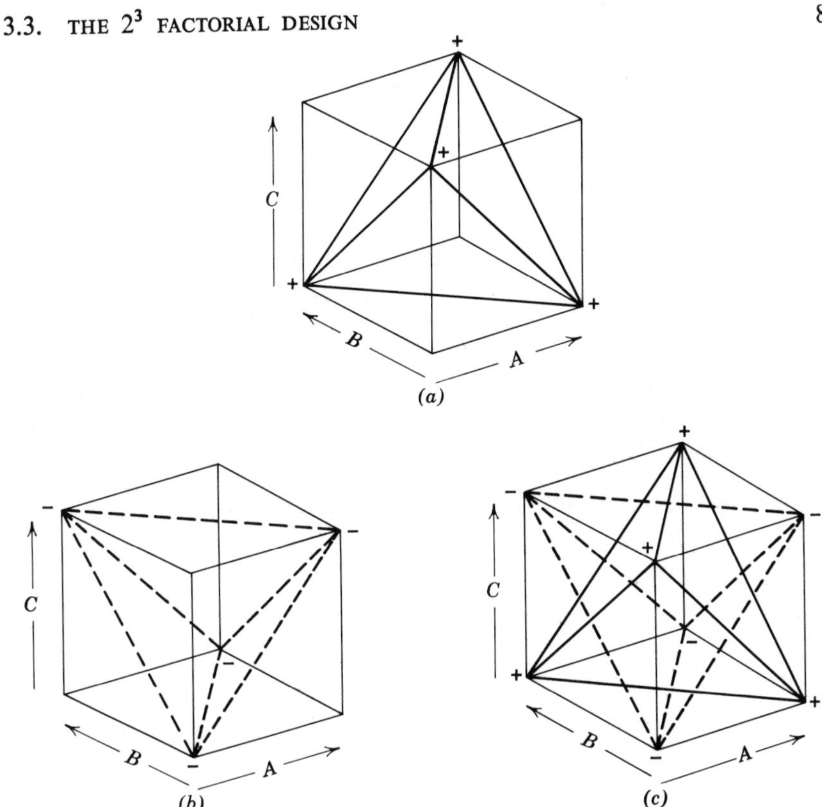

Figure 3.13. *Diagrammatic representation of the three-factor interaction ABC:* (*a*) *results having positive signs;* (*b*) *results having negative signs;* (*c*) *ABC contrast*

estimate the three main effects A, B, C and the three two-factor interactions AB, AC, BC. A further contrast which is of interest is the three-factor interaction ABC, which may be defined as one-half of the change in the AB interaction resulting from changing C.

From Figure 3.12*a* we see that we can make the calculations set out in Table 3.4.

It is easy to verify that exactly the same result would be obtained if we defined the ABC interaction as one-half the change in AC when B is changed or as one-half the change in BC when A is changed.

By considering how the various results have contributed to the calculation of the ABC interaction in Table 3.4, it will be found that this interaction can again be written as a contrast between two averages of four results.

$$
\begin{aligned}
ABC &= \tfrac{1}{4}(\bar{y}_2 + \bar{y}_3 + \bar{y}_5 + \bar{y}_8) - \tfrac{1}{4}(\bar{y}_1 + \bar{y}_4 + \bar{y}_6 + \bar{y}_7) \\
&= \tfrac{1}{4}(4.8 + 2.9 + 3.5 + 4.4) - \tfrac{1}{4}(3.9 + 3.6 + 2.8 + 5.1) \\
&= 3.90 - 3.85 \\
&= 0.05.
\end{aligned}
$$

Figure 3.13 shows the nature of this contrast. The observations $(\bar{y}_2, \bar{y}_3, \bar{y}_5, \bar{y}_8)$ and $(\bar{y}_1, \bar{y}_4, \bar{y}_6, \bar{y}_7)$ respectively fall at the four corners of two regular tetrahedra which fit inside the cube.

Three factor interactions are usually not of much practical importance themselves. We can, however, use the *ABC* contrast for a different purpose, specifically to arrange the design in "blocks" and so eliminate some of the extraneous variation. We discuss this application later.

Summary of Effects and Interactions in a 2^3 Factorial Design

For convenience we repeat here all the main effects and interactions we have already calculated.

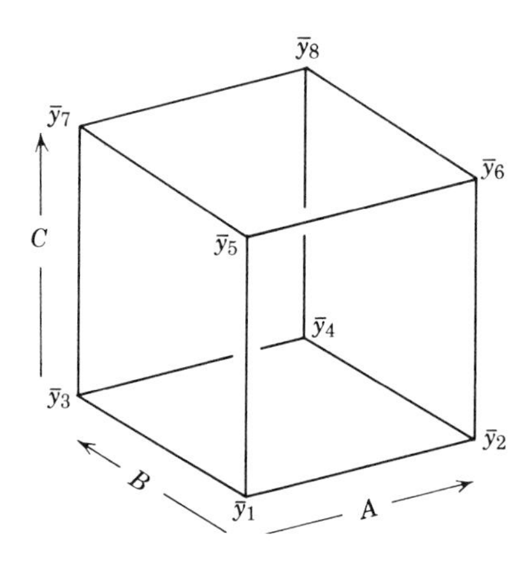

Main effects (see Figure 3.11):

$$
\begin{aligned}
A &= \tfrac{1}{4}(\bar{y}_2 + \bar{y}_4 + \bar{y}_6 + \bar{y}_8) - \tfrac{1}{4}(\bar{y}_1 + \bar{y}_3 + \bar{y}_5 + \bar{y}_7) = 1.20, \\
B &= \tfrac{1}{4}(\bar{y}_3 + \bar{y}_4 + \bar{y}_7 + \bar{y}_8) - \tfrac{1}{4}(\bar{y}_1 + \bar{y}_2 + \bar{y}_5 + \bar{y}_6) = -0.90, \\
C &= \tfrac{1}{4}(\bar{y}_5 + \bar{y}_6 + \bar{y}_7 + \bar{y}_8) - \tfrac{1}{4}(\bar{y}_1 + \bar{y}_2 + \bar{y}_3 + \bar{y}_4) = 0.15.
\end{aligned}
$$

Two-factor interactions (see Figure 3.12):

$$
AB = \tfrac{1}{4}(\bar{y}_1 + \bar{y}_4 + \bar{y}_5 + \bar{y}_8) - \tfrac{1}{4}(\bar{y}_2 + \bar{y}_3 + \bar{y}_6 + \bar{y}_7) = -0.05,
$$

$$AC = \tfrac{1}{4}(\bar{y}_1 + \bar{y}_3 + \bar{y}_6 + \bar{y}_8) - \tfrac{1}{4}(\bar{y}_2 + \bar{y}_4 + \bar{y}_5 + \bar{y}_7) = 0.40,$$

$$BC = \tfrac{1}{4}(\bar{y}_1 + \bar{y}_2 + \bar{y}_7 + \bar{y}_8) - \tfrac{1}{4}(\bar{y}_3 + \bar{y}_4 + \bar{y}_5 + \bar{y}_6) = 0.20.$$

Three-factor interaction (see Figure 3.13):

$$ABC = \tfrac{1}{4}(\bar{y}_2 + \bar{y}_3 + \bar{y}_5 + \bar{y}_8) - \tfrac{1}{4}(\bar{y}_1 + \bar{y}_4 + \bar{y}_6 + \bar{y}_7) = 0.05.$$

Note: A table of plusses and minuses to aid in the evaluation of all main effects and interactions in a 2^p design for $p = 2, 3, \ldots, 6$, is given by Beyer (1968, pp. 106–109).

Yates' Algorithm

The calculation of the main effects and interactions directly from the foregoing formulas can be tedious. A simpler routine method that can be applied to any 2^p design is due to Dr. Frank Yates. Its application to the 2^3 arrangement is illustrated in Table 3.5 for the by-product yield data.

Table 3.5. Calculation of effects for by-product data using Yates' algorithm

Row	Factor Combination A	B	C	Average Response	Columns 1	2	3	Divisor	Effect, Col. 3/Divisor	Nature of Effect
1	−	−	−	3.9	8.7	15.2	31.0	8	3.875	Mean
2	+	−	−	4.8	6.5	15.8	4.8	4	1.20	*A*
3	−	+	−	2.9	8.6	1.6	−3.6	4	−0.90	*B*
4	+	+	−	3.6	7.2	3.2	−0.2	4	−0.05	*AB*
5	−	−	+	3.5	0.9	−2.2	0.6	4	0.15	*C*
6	+	−	+	5.1	0.7	−1.4	1.6	4	0.40	*AC*
7	−	+	+	2.8	1.6	−0.2	0.8	4	0.20	*BC*
8	+	+	+	4.4	1.6	0.0	0.2	4	0.05	*ABC*
S of *S* check				125.08			1000.64			

The eight sets of conditions for the variables *A*, *B*, and *C* are first written down using the symbol minus to indicate the low level and the symbol plus to indicate the high level. For a qualitative factor such as *catalyst supplier*, we arbitrarily assign one supplier the minus symbol and the other the plus symbol. The effect as then calculated will be the average difference

$$(\text{supplier labeled } +) - (\text{supplier labeled } -).$$

To employ the algorithm we must first ensure that the results are written down in a particular order. To achieve this, the A column is written down following the sequence minus, plus, minus, plus, and so on, until we have eight symbols corresponding to the eight runs of the table. The column for factor B is then written down with *two* minuses followed by *two* plusses until we have eight symbols again. Finally, the C column has *four* minuses followed by *four* plusses. The average responses associated with these conditions are next entered. The ordering thus achieved is called *standard order* and is precisely the one that we have employed previously in discussing the design. The eight rows of the table are marked off by horizontal lines to form four successive pairs in the manner shown. The first four numbers in "column 1" are the sums of the pairs. Thus $8.7 = 3.9 + 4.8$, $6.5 = 2.9 + 3.6$, $8.6 = 3.5 + 5.1$, $7.2 = 2.8 + 4.4$. The last four numbers in column 1 are differences of the pairs. Thus $0.9 = 4.8 - 3.9$, $0.7 = 3.6 - 2.9$, $1.6 = 5.1 - 3.5$, $1.6 = 4.4 - 2.8$. Column 2 is now derived by performing on the entries of column 1 the same operations that we used to form column 1 from the average response column. Thus $15.2 = 8.7 + 6.5$, $15.8 = 8.6 + 7.2, \ldots, -2.2 = 6.5 - 8.7$, and so on. Finally, column 3 is calculated in the same way from column 2. The entries in column 3, divided by their divisors, which are 8 for the first entry and 4 for the remainder, provide the *effects*. These effects also occur in "standard order." The identity of the effects can be obtained by writing in the last column the letters which coincide with plus signs in the corresponding factor combination. Thus in row 2 the factor combination shown is $(+\ -\ -)$ and the corresponding effect indicated in the last column is A. The factor combination in row 4 is $(++-)$, and the corresponding effect indicated in the last column is AB and so on. The effect in the first row corresponding to $(---)$ is the mean of the eight responses. The effects thus identified will be found to be identical with those obtained previously.

The Yates procedure can be used in the analysis of any 2^p type factorial design. In particular, it can be used in the analysis of the 2^2 design. The calculations for this two-variable design are so simple, however, that probably little is gained in this case. In Chapter 4 alternative worksheet layouts are shown using (*a*) the direct calculation and (*b*) the Yates procedure.

Checking the Yates Algorithm

The most reliable verification of the Yates calculation is supplied by what is called the *sum of squares check* or *S of S check*. Suppose the sum of squares of the elements in the "data column" (which contains the original average responses or the original average responses after some convenient constant has been subtracted from each) is S. Then, if the calculations are correctly made, the sum of squares of the elements of columns 1, 2, and 3 are exactly $2S$, $4S$, and $8S$ respectively (it can be shown). In practice we usually proceed by first calculating the sums of squares of the data column and of column 3 only. If the latter is exactly eight times the former, it can be assumed that the calculation is correct. Otherwise, the sum of squares for column 2 and, if necessary, column 1 can be checked to find the column in which the error is located. In the foregoing sample calculation, $1000.64 = 8 \times 125.08$ and the check is satisfied.

Standard Errors of Effects for Two-level Factorials

We have discussed in some detail the 2^2 and the 2^3 factorial designs. From this discussion the reader will appreciate that, if we perform n cycles of a 2^p design in p factors, each main effect and interaction will be a contrast between the average of one half of the observations and the average of the other half.

If σ^2 is the variance of the individual observations supposed independently distributed, then the variance of an average of $\frac{1}{2}(n \times 2^p)$ observations is $2\sigma^2/n2^p$. Each main effect and interaction is the difference of two such independent differences and, therefore, has variance

$$\frac{2\sigma^2}{n2^p} + \frac{2\sigma^2}{n2^p} = \frac{4\sigma^2}{n2^p}.$$

The standard error for each effect will be obtained by taking the square root of this quantity and substituting the estimate s for the standard deviation σ, yielding the results set out in Table 3.6.

Table 3.6. Variances and standard errors for main effects and interactions estimated from a 2^p factorial design after n cycles

Design	2^p	2^2	2^3
Variance (effect)	$\frac{4\sigma^2}{n2^p}$	$\frac{\sigma^2}{n}$	$\frac{\sigma^2}{2n}$
S. E. (effect)	$\frac{2s}{\sqrt{n2^p}}$	$\frac{s}{\sqrt{n}}$	$\frac{s}{\sqrt{2n}}$

For illustration we return to the by-product example. We show later that for these data the estimated standard deviation s after three cycles of a 2^3 design is $s = 0.38$. The 2 S.E. limits for the effects are therefore provided by

$$\text{effect} \pm 2\left(\frac{s}{\sqrt{2n}}\right),$$

where $n = 3$, $s = 0.38$. Thus the 2 S.E. limits for each main effect and interaction are given by

$$\text{effect} \pm 0.31.$$

Interpretation of the Effects for the 2^3 Factorial

Using the by-product yield data for illustration, we now consider how the various statistics we have discussed may be brought together in the interpretation of results from an EVOP scheme using a 2^3 design.

Table 3.7 shows the results as they would be set out on the information board. The by-product averages after three cycles of operation are shown at the corners of the cube. The estimated main effects and interactions with their 2 S.E. limits are given below the cube.

In considering this display, we apply precisely those principles discussed already for the 2^2 design. Attention is focused on those effects that are

Table 3.7. Information board for by-product yields after three cycles

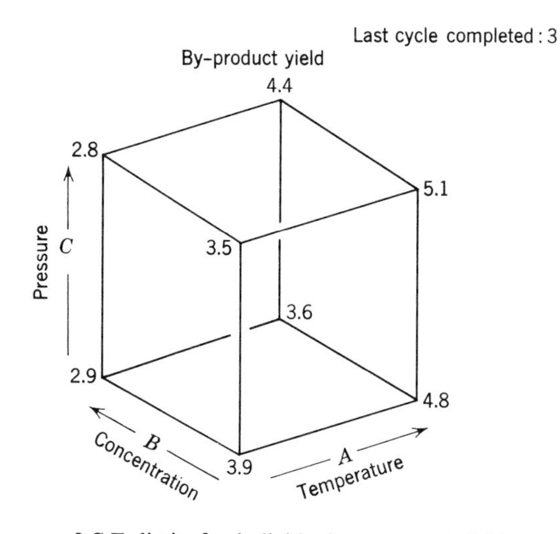

2 S.E. limits for individual averages: ± 0.44

Temperature	A	1.20 ± 0.31
Concentration	B	-0.90 ± 0.31
Pressure	C	0.15 ± 0.31
Temperature × concentration	AB	-0.05 ± 0.31
Temperature × pressure	AC	0.40 ± 0.31
Concentration × pressure	BC	0.20 ± 0.31
	ABC	0.05 ± 0.31

reasonably large in magnitude compared with their standard errors. These are the main effect of temperature $A = 1.20$, the main effect of concentration $B = -0.90$, and the temperature-pressure interaction $AC = 0.40$.

Before interpreting the main effect of a given factor, we must look to see whether there are any appreciable interactions that involve this factor. If there are, then the interacting factors are interpreted together rather than separately. We see in the case of concentration B that, although there is a large main effect $B = -0.90$, the interactions $AB = -0.05$ and $BC = +0.20$ involving B are small, both in relative magnitude and in relation to their standard errors. We may conclude therefore that "a change in by-product yield of -0.9 ± 0.3 is associated with increased concentration, and this effect does not seem to depend on which of the tested combinations of temperature and pressure are used."

By contrast, although there is a large main effect $A = 1.20$ associated with temperature, there is also an appreciable interaction $AC = -0.40$ of temperature with pressure. The effect of temperature and pressure must, therefore, be assessed jointly rather than individually. This can most easily be done by setting out a 2×2 table of averages for the factors concerned. In this instance, averaging over concentration, we have:

Yields of by-product averaged over concentration *

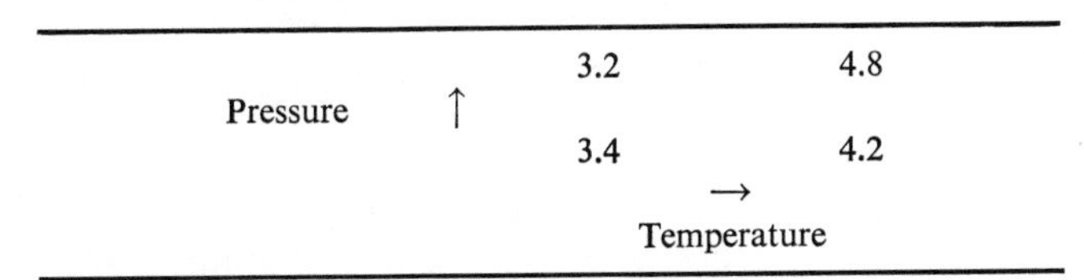

Pressure ↑	3.2	4.8
	3.4	4.2
	Temperature →	

* Here and elsewhere we have rounded a five in the last place up if the previous digit is odd and down otherwise. In other words, we have always rounded a five to an even digit.

The table summarizes what the data have to tell us about the temperature and pressure effects and, for example, may be interpreted as follows. Higher temperature is associated with increased by-product (see two-way table for magnitude of effects) irrespective of which level of pressure is used. At the higher temperature, increased pressure probably further increases by-product yield.

In practice there would usually be other responses to consider which might constrain possible action. If there were no such constraints, then in the furtherance of reducing by-product yield we should explore conditions of lower temperature and high concentration, keeping the median level of pressure close to its previous value.

As before, to assist in the visualization of reasonably well-established

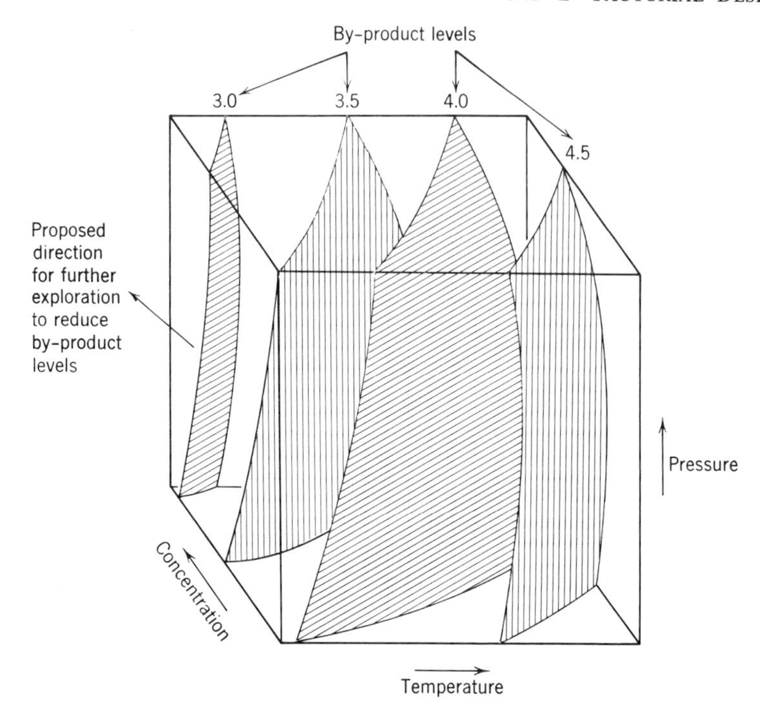

Figure 3.14. Approximate by-product contours drawn by eye interpolation among the points with proposed direction for further exploration indicated.

effects, rough and informal contour sketching is helpful if used with discretion. Figure 3.14 shows such a set of eye-drawn contours for the present example in relation to the proposed direction for further exploration. As with two-variable problems, approximate contour plots of this kind can be very helpful when we have to consider several responses simultaneously.

An Estimate of the Standard Deviation

To compute the standard errors of the effects, we need an estimate s of the standard deviation σ of the individual observations.

To illustrate the way in which this is obtained, we once more employ the by-product data. In our previous analysis of these data we had been concerned only with the averages obtained after three cycles and rounded to one decimal. The data for the individual runs are given in Table 3.8.

The reader will notice that, as soon as two or more cycles have been completed, we have information on how closely results at each set of process conditions are reproduced in repeated trials. This allows us to obtain an estimate s of the standard deviation σ of a single observation from the

Table 3.8 By-product data for three repetitions of a 2^3 design

Process Conditions	Cycle 1	Cycle 2	Cycle 3	Averages After Three Cycles
1	3.9	4.2	3.6	3.90
2	4.3	5.1	5.1	4.83
3	3.4	2.5	2.7	2.87
4	3.8	3.3	3.7	3.60
5	3.7	3.3	3.6	3.53
6	5.2	5.5	4.6	5.10
7	2.9	2.8	2.7	2.80
8	4.7	4.4	4.0	4.37

observed reproducibility of the results. In the specific set of data above we see that, after the third cycle, there are three repetitions at each set of process conditions. We could, therefore, obtain an estimate of variance based on two degrees of freedom from each such set. On the assumption that the true variances are equal at the various sets of conditions, these estimates could be averaged to produce an estimate of the variance σ^2 having $8 \times 2 = 16$ degrees of freedom.

This analysis would be entirely appropriate if we could be sure that there were no systematic differences from cycle to cycle. However, changes in raw material, shifts in ambient temperature, and other causes could produce such differences. To avoid overestimation of the errors, we should eliminate such effects by removing over-all differences between the column means of the table as well as differences between row means. The orthodox way of doing this is called a *two-way analysis of variance*. This requires much summing of squared quantities and is a procedure not well suited to the conditions under which EVOP is usually applied. We substitute a simplified method that employs the range. A justification of the procedure is given in Appendix 1.

Now, when only a single cycle of runs has been completed, we cannot make any estimate of the reliability of the data themselves; but as soon as a second cycle has been run such an estimate can be made by comparing the new results with the old. The by-product data for the first two cycles, with differences and averages, are given in Table 3.9. Obviously the eight differences cycle 1 $-$ cycle 2 supply information about σ which is supposedly the same at each of the eight process conditions.

We have seen earlier that an estimate of the standard deviation of a set of numbers is produced by multiplying the range by a suitable factor w.

Table 3.9. ***Calculation to obtain an estimate s_2 of σ after two cycles***

Process Conditions	Cycle 1	Cycle 2	Difference, Cycle 1 − Cycle 2	Average, (Cycle 1 + Cycle 2)/2
1	3.9	4.2	−0.3	4.05
2	4.3	5.1	−0.8 [1]	4.70
3	3.4	2.5	0.9 [1]	2.95
4	3.8	3.3	0.5	3.55
5	3.7	3.3	0.4	3.50
6	5.2	5.5	−0.3	5.35
7	2.9	2.8	0.1	2.85
8	4.7	4.4	0.3	4.55

[1] Highest and lowest differences.

When there are eight observations, this factor (see Table 2.3, Chapter 2) is $w_8 = 0.351$. The range for the eight differences in this case is $R = 0.9 - (-0.8) = 1.7$, so that the estimated *standard deviation of the differences* is $Rw_8 = 1.7 \times 0.351 = 0.597$.

Now, in Chapter 2, Section 2.7, we have seen that, on the assumption of independence, the standard deviation of the difference of two observations (each having standard deviation σ) is $\sqrt{2}\,\sigma$. By dividing the estimate 0.597 by $\sqrt{2}$, we, therefore, obtain an estimate of σ:

$$s = \frac{0.351 \times 1.7}{1.414} = 0.42.$$

Denoting the range R calculated after two cycles by R_2 and the corresponding estimate of σ by s_2, we have

$$s_2 = \sqrt{\tfrac{1}{2}}\, w_8 R_2.$$

As soon as the results for the third cycle are available, more comparisons can be made and a more accurate estimate of σ can be obtained. This is done as follows. At the completion of the third cycle we have available the averages of the first two cycles representing the "old data" and the new set of eight results. The differences supply further information about the standard deviation.

The range of the differences, which we denote by R_3, is (see Table 3.10)

$$R_3 = 0.75 - (-0.40) = 1.15.$$

A new estimate of the standard deviation is now calculated from

$$s_3 = \sqrt{\tfrac{2}{3}}\, w_8 R_3 = \frac{1.414}{1.732} \times 0.351 \times 1.15 = 0.33.$$

Table 3.10. Calculation to obtain a second estimate s_3 of σ after three cycles

Averages for Cycles 1 and 2	Cycle 3	Differences
4.05	3.6	0.45
4.70	5.1	−0.40 [1]
2.95	2.7	0.25
3.55	3.7	−0.15
3.50	3.6	−0.10
5.35	4.6	0.75 [1]
2.85	2.7	0.15
4.55	4.0	0.55

[1] Highest and lowest differences.

This new estimate of σ may now be averaged with the previous estimate to give

$$s = \tfrac{1}{2}(0.42 + 0.33) = 0.38,$$

which we have used in the calculation of the standard errors after three cycles earlier in this section. After a further cycle is completed, we can compare the averages from the first three cycles with results from the fourth cycle. The range of the differences R_4 yields

$$s_4 = \sqrt{\tfrac{3}{4}}\, w_8 R_4,$$

which may be pooled with the previous two estimates.

We show in Appendix 1 that, in general, if there are k sets of process conditions ($k = 8$ for our example) and if n cycles have been completed, then the differences of the averages of the first $n - 1$ cycles and the new data in the nth cycle yield an estimate

$$s_n = \left(\frac{n-1}{n}\right)^{1/2} w_k R_n,$$

where w_k is the range factor given in Table 2.3 of Chapter 2. In practice, it is convenient to tabulate the factor $[(n - 1)/n]^{1/2} w_k$ once and for all. We call this factor $f_{k,n}$.

The table of $f_{k,n}$ for $n = 2, 3, \ldots, 20$ and $k = 2, 3, \ldots, 10$ is given in Table 3.11 and also in Appendix 1.

We now illustrate the use of this table; in our example we have $k = 8$. For $n = 2$, $f_{8,2} = 0.25$, so that

$$s_2 = 0.25 \times 1.7 = 0.42$$

Table 3.11. A table of values of $f_{k,n}$

No. of Cycles, n	Number of runs in the Block, k								
	2	3	4	5	6	7	8	9	10
2	0.63	0.42	0.34	0.30	0.28	0.26	0.25	0.24	0.23
3	0.72	0.48	0.40	0.35	0.32	0.30	0.29	0.27	0.26
4	0.77	0.51	0.42	0.37	0.34	0.32	0.30	0.29	0.28
5	0.79	0.53	0.43	0.38	0.35	0.33	0.31	0.30	0.29
6	0.81	0.54	0.44	0.39	0.36	0.34	0.32	0.31	0.30
7	0.82	0.55	0.45	0.40	0.37	0.34	0.33	0.31	0.30
8	0.83	0.55	0.45	0.40	0.37	0.35	0.33	0.31	0.30
9	0.84	0.56	0.46	0.40	0.37	0.35	0.33	0.32	0.31
10	0.84	0.56	0.46	0.41	0.37	0.35	0.33	0.32	0.31
15	0.86	0.57	0.47	0.42	0.38	0.36	0.34	0.33	0.31
20	0.86	0.58	0.47	0.42	0.38	0.36	0.34	0.33	0.32

as before. For $n = 3, f_{8,3} = 0.29$, so that

$$s_3 = 0.29 \times 1.15 = 0.33$$

as before.

It should be noted that this method of calculation does eliminate differences between cycles. To see this, suppose an arbitrary quantity x were added to the averages in Table 3.10 and an arbitrary quantity y to the results of cycle 3. Then we should have the following table:

Averages for Cycles 1 and 2	Cycle 3	Differences
$4.05 + x$	$3.6 + y$	$0.45 + x - y$
$4.70 + x$	$5.1 + y$	$-0.40 + x - y$ [1]
$2.95 + x$	$2.7 + y$	$0.25 + x - y$
$3.55 + x$	$3.7 + y$	$-0.15 + x - y$
$3.50 + x$	$3.6 + y$	$-0.10 + x - y$
$5.35 + x$	$4.6 + y$	$0.75 + x - y$ [1]
$2.85 + x$	$2.7 + y$	$0.15 + x - y$
$4.55 + x$	$4.0 + y$	$0.55 + x - y$

[1] Highest and lowest differences.

The range $0.75 + x - y - (-0.40 + x - y) = 1.15$ remains unchanged.

A further attractive feature of the method is its insensitivity to rounding error. Such rounding as is done will be directly transmitted to the range and will not be magnified.

The method has, therefore, a number of advantages.

1. It requires only addition, subtraction, and multiplication and no summing of squares.
2. For k in the range of the table, the efficiency of the method is high. (The variance of the estimate s will be only slightly larger than that which would be obtained from the estimate provided by a full analysis of variance.)
3. Only the quantities "old averages" and "new results," which at any given stage are immediately available, are used in the calculation.
4. Rounding errors are controlled.
5. Cycle differences are eliminated automatically.

A procedure similar to the one illustrated is, of course, used in the two-factor case.

3.4. DIVIDING THE 2^3 FACTORIAL DESIGN INTO TWO BLOCKS

The use of the 2^3 design allows us to study the main effects and interactions among three variables with the same precision as would be obtained from the same number of runs applied to only two factors. In a sense, therefore, we obtain the information about the third factor and its interactions as a "bonus." The same argument suggests that it would be even more advantageous to study four or more factors simultaneously. Designs involving more factors are, in fact, of great value for special "one-shot" studies where some particular problem has to be examined for a limited period of time with special process supervision. The authors' experience has been, however, that the study of three variables simultaneously usually represents the practical limit under the circumstances of a genuine EVOP program. These circumstances, it will be remembered, demand that the program must be run by the process operators themselves as a continuing routine which can be tolerated indefinitely and does not disrupt production.

The argument that three variables can be examined as economically as two assumes that the error of an individual run is the same for the smaller 2^2 design as it is for the larger 2^3 design. This assumption is less innocuous than it appears at first sight. It will be recalled that, whether we are running two variables or three, systematic differences between cycles are automatically eliminated. The error s that we estimate and which determines our 2 S.E. limits is an estimate of the variation that would occur *within* a cycle if no deliberate changes were made. We would expect that more opportunity for variation would occur in a cycle of $2^3 = 8$ runs than in a cycle of $2^2 = 4$ runs. Fortunately, we can gain all the advantages of the smaller cycle size with the larger 2^3 design by running it in two halves, that is, in two *blocks* of four runs each. This can be done in such a way that all the effects and interactions of interest are calculated from comparisons made *within a block*.

The idea may be understood by imagining the situation where we can make four of the eight runs of the 2^3 factorial on one day, and four runs on the next day. We can pose the question: "How should we arrange the eight runs so that possible systematic day-to-day differences will not affect the comparisons of principal interest?" The comparisons of principal interest are the main effects and the two-factor interactions. The three-factor interaction contrast ABC will usually be small and of lesser interest, so we use it to accommodate the day-to-day difference. Now (Fig. 3.13) the interaction is a contrast between runs 2, 3, 5, and 8 and runs 1, 4, 6, and 7. Our blocking plan, then, consists of making runs 2, 3, 5, and 8 on one day, and runs 1, 4, 6, and 7 on the other. We then have

$$ABC = \tfrac{1}{4}(y_2 + y_3 + y_5 + y_8) - \tfrac{1}{4}(y_1 + y_4 + y_6 + y_7),$$
$$\text{day 1} - \text{day 2} = \tfrac{1}{4}(y_2 + y_3 + y_5 + y_8) - \tfrac{1}{4}(y_1 + y_4 + y_6 + y_7).$$

In other words, the contrast which measures the ABC effect also measures the day 1 $-$ day 2 effect. These two effects are said to be *confounded*, that is, confused with each other. If we attempt to estimate the ABC interaction, what we shall really obtain is the effect ABC *plus* the contrast between day 1 and day 2, and it will be impossible to know how much of the effect is produced by the interaction and how much by the day-to-day difference. By this sacrifice of information about the three-factor interaction, however, *we can keep all other effects free from day-to-day differences.* This can be seen in the following way.

Suppose the values of the observations on day 1 are, on the average, x units higher than those on day 2. Then, if we let $y_1, y_2, \ldots, y_8$ denote the results that would have been obtained if all the runs had been conducted on day 2, the results actually found would be:

$$\begin{aligned} y_1' &= y_1, & y_5' &= y_5 + x, \\ y_2' &= y_2 + x, & y_6' &= y_6, \\ y_3' &= y_3 + x, & y_7' &= y_7, \\ y_4' &= y_4, & y_8' &= y_8 + x. \end{aligned}$$

Now suppose we calculated the main effect of A from these results. We should have

$$\tfrac{1}{4}(y_2' + y_4' + y_6' + y_8') - \tfrac{1}{4}(y_1' + y_3' + y_5' + y_7')$$
$$= \tfrac{1}{4}[(y_2 + x) + y_4 + y_6 + (y_8 + x)] - \tfrac{1}{4}[y_1 + (y_3 + x) + (y_5 + x) + y_7].$$

We see that two of the x's appear in the first parentheses with a plus sign and two of the x's appear in the second parentheses with a minus sign. The day-to-day effect thus cancels out. This happens not only for the A effect but for all the other main effects and two-factor interactions, in fact for every contrast except that for the ABC interaction, where all four x's will appear in the first bracket. What is happening is made clearer by looking at Figure 3.15, where the additional quantity $+x$ is shown associated with the four runs numbered 2, 3, 5, and 8. Reference to Figure 3.11 will show that two x's lie on each of the pairs of faces of the cube which provide the main effect contrasts. Moreover, reference to Figure 3.12 will show that two x's lie on each of the pairs of diagonal planes of the cube which provide the two-factor interaction contrasts.

This rather remarkable property arises from what is called the *orthogonality* of the design. It will be found that we could have used any two-factor interaction or any main effect as a basis for confounding without affecting the other estimates. It is obviously most sensible, however, to sacrifice information on what is expected to be the least important effect, namely, ABC.

Although our discussion has been in terms of single observations from a single cycle, similar results and remarks apply if the averages obtained from n cycles of operation are used.

In the foregoing, the blocks represented two different days of running. Often, a block is a subset of experimental material which is expected to be more homogeneous than the complete set. Thus variations arising from slight differences in batches of raw material could be eliminated if four EVOP runs could be made within each batch. In this case the blocks would be "batches of raw material." When periods of time are associated with blocks, the period need not, of course, be a single day. If four batches could be made in one shift, a shift could be a block. Even when the time taken to make four runs is some odd period such as 19 hours, it would

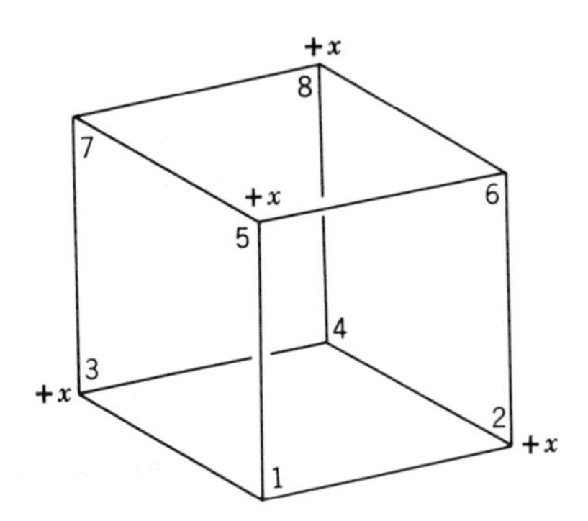

Figure 3.15. Additional responses on day 1.

usually still be worthwhile to block the eight runs into two sets of four. A group of runs made close together in time will normally show greater similarity than runs which cover a longer period. In general, then, by introducing the additional refinement of blocking, we can study three variables in an EVOP scheme without introducing additional variation beyond that to be expected for a two-variable scheme.

The rationale for the choice of the systematic pattern for use of the 2^2 design when randomization was not possible will now be clear. The sequence order

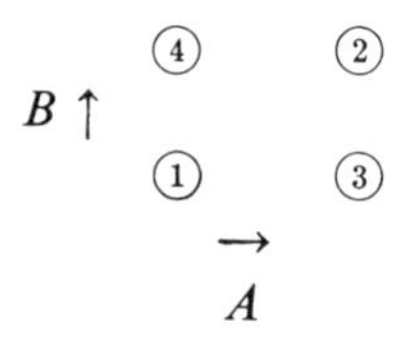

ensures that systematic differences between the first pair of runs (1, 2) and the second pair (3, 4) is associated with the interaction AB rather than with the main effect A or B. In practice, this will mean that the interaction has a precision slightly less than that implied by the standard error estimate, whereas the main effects have a precision slightly greater.

Inclusion of Reference Conditions With 2^3 Designs

As with the two-variable scheme, it is comforting and often helpful to revert periodically to a set of reference conditions. Exactly as before, these would usually be conditions thought to be currently best. Also as before, the conditions could correspond to one of the design points, to a point *outside* the design or to a point *inside* the design. Also we may revert to the reference conditions once per cycle, or more often, or less often. In any case, the *change in mean* comparison, i.e. (average over all conditions run in the EVOP cycle) − (average at reference condition), may be calculated and will estimate the temporary cost, if any, of information gathering in terms of material, quality, or money.

In the special case where the reference conditions are at the *center* of the factorial design, the change in mean will, in addition, provide a measure of the over-all curvature of the response surface.

A 2^3 Design With Two Center Points Run in Two Blocks

One design of this kind which is of particular interest is that shown in Figure 3.16 in which the eight runs of the 2^3 design are split into two blocks of four runs, as previously described, and a center point is associated with each block. We emphasize again that this is not the only arrangement that can be used, but is one that has sometimes been found convenient in practice; we shall discuss it in Chapter 5 as a model arrangement to illustrate the use of calculation worksheets.

As with the two-factor design when randomization cannot be conveniently applied, the design has been run in a particular sequence and the runs have been renumbered accordingly. This *sequence numbering* is used in Figure 3.16 and in Chapter 5. The center conditions in the first block have been labeled 0 (zero) and in the second Ø (referred to as zero slash).

So far as the effect of possible serial correlation between errors is concerned, very little is lost by using this systematic pattern instead of a random pattern. Although it is true that the existence of correlations between successive observations can invalidate *certain* types of statistical analysis,

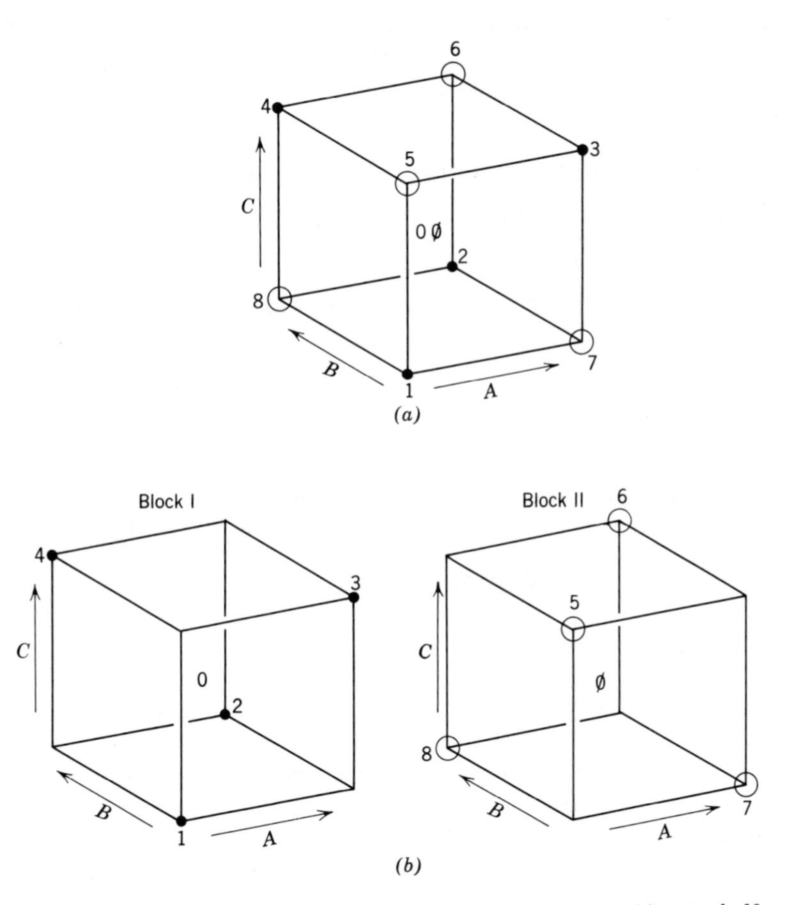

Figure 3.16. *A model three-variable EVOP scheme with center points: (a) a single 10-point cycle in two blocks (sequence numbering); (b) the two blocks of the design (sequence numbering).*

it has been shown [Box (1954*a*)] that comparisons of the type we make here are only slightly affected.

From the design of Figure 3.16, two measures of the change in mean may be calculated, one from each block of the design. For block I, in terms of single observations,

$$\text{change in mean} = \tfrac{1}{5}(y_1 + y_2 + y_3 + y_4 - 4y_0).$$

For block II,

$$\text{change in mean} = \tfrac{1}{5}(y_5 + y_6 + y_7 + y_8 - 4y_\emptyset).$$

The average,

$$\tfrac{1}{10}(y_1 + y_2 + y_3 + y_4 - 4y_0 + y_5 + y_6 + y_7 + y_8 - 4y_\emptyset),$$

provides a measure of the *over-all change in mean* free from block effects. It has precisely the same interpretation as in the two-variable EVOP and, as always, is equal to the phase mean minus the mean at the reference conditions.

Two Standard Error Limits for the Over-all Change in Mean

After n cycles of operation the over-all change in mean would be

$$-\tfrac{4}{10}\bar{y}_0 + \tfrac{1}{10}\bar{y}_1 + \tfrac{1}{10}\bar{y}_2 + \tfrac{1}{10}\bar{y}_3 + \tfrac{1}{10}\bar{y}_4 - \tfrac{4}{10}\bar{y}_\emptyset + \tfrac{1}{10}\bar{y}_5 + \tfrac{1}{10}\bar{y}_6 + \tfrac{1}{10}\bar{y}_7 + \tfrac{1}{10}\bar{y}_8.$$

Hence, using (2.7.4), the variance of the over-all change in mean is

$$\left(\frac{16}{100} + \frac{1}{100} + \frac{1}{100} + \frac{1}{100} + \frac{1}{100} + \frac{16}{100} + \frac{1}{100} + \frac{1}{100} + \frac{1}{100} + \frac{1}{100}\right)\frac{\sigma^2}{n}$$
$$= 0.4\,\frac{\sigma^2}{n}.$$

It follows that the standard deviation of the over-all change in mean is $0.632\sigma/\sqrt{n}$. If the estimate of σ is s, the 2 S.E. limits will be

$$\text{Over-all change in mean} \pm 2(0.632)\frac{s}{\sqrt{n}}.$$

3.5. SUMMARY

This chapter together with the foregoing chapter has provided an introduction to those parts of statistical theory, design, and analysis necessary in the application of EVOP. The two chapters that follow show how these principles are applied to numerical examples and also introduce special worksheets for model designs illustrating a systematic and simplified calculation procedure that may be employed. By applying the general principles outlined in Chapters 2 and 3, the investigator should have no difficulty in adapting these schemes to special circumstances that will arise from time to time.

CHAPTER 4

Worksheets for Two-variable *EVOP* Programs

4.1 INTRODUCTION

In practice EVOP is run in the factory by process personnel. Consequently, it must be reduced to a routine which (*a*) can be easily handled without jeopardizing the efficient operation of the process and (*b*) can be carried out by persons who have no special statistical training.

The basic elements in organizing an EVOP program usually are (*a*) running in sequence a pattern of conditions which are variants of the currently best-known process and (*b*) recording the results in a readily understood form which points the way to appropriate action.

We have seen that a convenient pattern of variants consists of a simple two-level factorial design, often with an added point that corresponds to the currently best known conditions. We have also seen that efficient feedback may be achieved by setting up an information board (such as Figure 4.1) to display a readily understood analysis of the results. It is important that some specific person be assigned the task of keeping the information board up to date. This task must be promptly and conscientiously performed.* Tardy or inaccurate information can seriously prejudice success. Now it will be recalled that the numbers which appear on the information board are not the raw data but various averages, effect contrasts, and

* The worksheet calculations we describe can be performed reasonably quickly with pencil and paper only and allow EVOP to be carried out under the most primitive conditions. In making the calculations, a slide rule or simple desk calculator is helpful, but not essential. On some locations, in the more highly industrialized countries, electronic computational facilities are available. Such facilities are by no means necessary to the success of EVOP, of course. However, if they are available, they should certainly be used to ensure promptness and accuracy in the calculations.

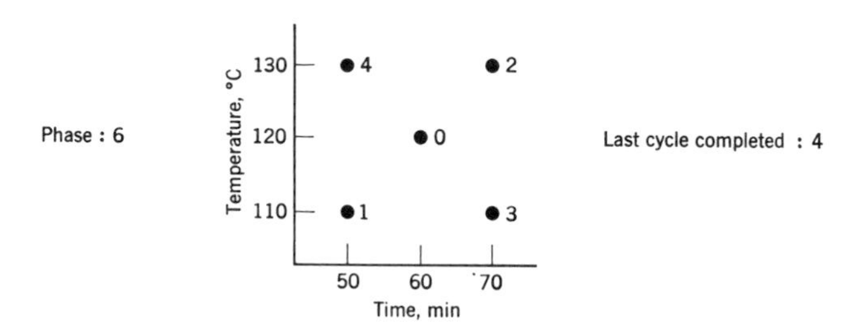

		Yield, %	Other Responses . . .
Requirement		Maximize	
Process averages		Temp ↑ 61.9 64.6 / 62.2 / 63.4 65.7 → Time	. . .
2 S. E. limits for averages		±2.3	. . .
Effects with 2 S. E. limits	Phase mean	63.6	
	Time	2.5 ± 2.3	
	Temp.	−1.3 ± 2.3	
	$t \times T$	0.2 ± 2.3	. . .
	Change in mean	1.4 ± 2.0	
Estimated standard deviation s		2.3	. . .
Prior estimate of standard deviation σ		1.8	. . .

Figure 4.1. *Part of an information board after four cycles.*

2 S. E. limits designed to provide understanding of the data. The object of the worksheets to be introduced is to reduce, to a simple routine, the calculation of the various quantities needed for display on the information board. It will be understood that the calculation worksheets are kept for reference but are not normally themselves displayed. As previously explained, appraisal of the analysis exhibited on the information board is made on a day-to-day basis by the process superintendent, with the periodic help—perhaps at monthly intervals—of the EVOP committee.

Before describing the worksheets, it may be helpful to remind the reader of the terminology we shall use.

Cycle and Phase. A single performance of a complete set of operating conditions is called a *cycle*, and the repeated running through of a cycle of operating conditions is called a *phase*. A new phase begins as soon as we commence to explore some new set of conditions.

Run. The observation entries on the calculation worksheets are the values obtained from individual runs. It will be remembered that what we here call a run corresponds to a period of operation in which the process conditions are maintained at some fixed levels. In the case of a batch process this could mean a single batch providing a single observation or a series of batches all run at the same operating conditions leading to an average observation which represents the run. In a continuous process a run would mean a suitable period of operation at fixed conditions. The period would normally be long enough to allow the process to "line out" at the new conditions and for adequate data to be taken. In some processes in which the inertia of the system was high the time to equilibrium might be very long, amounting to hours or, in rare cases, even to days. In other cases in which the inertia of the system was low, for example, in reactions between gases, the time to equilibrium could be much shorter.

Calculation and Use of Standard Deviation Entries

The calculation procedure described in detail for two-variable schemes in this chapter, and for three-variable schemes in Chapter 5, uses the range to provide an easily calculated estimate of the standard deviation σ and automatically removes block differences.

It will usually be found that the standard deviation does not change greatly from one phase of EVOP to another although, with the passage of time, there may be a tendency for variation to decrease to some extent because factors which have previously caused variation are tied down at their best levels as a result of previous EVOP phases. Once a reasonably good pooled estimate of the standard deviation σ has been obtained from the first few phases, this estimate is sometimes employed from then on unless checks indicate that it is no longer relevant.

In the routine we illustrate here, however, the following convention is employed when the 2^2 design is used. After the first and second cycles of each new phase, a *prior estimate of* σ obtained from previous phases is employed. With the completion of the third cycle, the use of the prior estimate is discontinued and the standard deviation is estimated from the data of the current phase.

The reasoning here is as follows. Usually the standard deviation has not changed very much from one phase to another, but occasionally slight changes do occur because of the nature of the factors which are being varied. After the first cycle of the new phase, we will have no information about σ from the current data. After two cycles an estimate will be possible but it will be very unreliable. With the completion of the third cycle an estimate of moderate accuracy is possible and, at that point, it may be profitable to use it. The procedure is sensible although obviously somewhat arbitrary. It can be deviated from if special circumstances suggest this to be desirable.

It will now be clear why we have ignored the uncertainty caused by sampling variation in our estimate of σ. The prior estimate, which we employ at the beginning of each new phase, will usually be based on a large number of degrees of freedom, and its sampling variation will therefore be small. After completion of the third cycle an estimate of σ known to be appropriate to the new phase will have been obtained, and by this time it will make little difference if the sampling variation in this new estimate is ignored.

At the very beginning of an EVOP program, there will be no prior estimate from previous phases. To get started, an estimate of the standard deviation calculated from plant records is used. Such an estimate may lead to conservative action, because a frequent consequence of introducing EVOP is to produce more careful operation of the process and a reduction in the standard deviation.

Inclusion of Reference Conditions

The calculation will differ slightly depending on whether or not one of the sets of conditions studied is a *reference condition*. If it is, then we should normally calculate the *change in mean* in addition to the other effects, otherwise not. Again, if a reference condition is included, then it will make some difference whether this condition is one of the factorial runs or whether it is an additional run.

4.2. WORKSHEETS FOR A 2^2 FACTORIAL DESIGN WITH ADDED REFERENCE CONDITIONS

We first demonstrate the use of calculation worksheets for a 2^2 design with an added reference condition. For illustration we show, in Figures 4.2, 4.3, and 4.4, calculations made in the sixth phase of an EVOP scheme applied to a certain batch process. In this process the effect of varying reaction time and temperature were being studied, using a 2^2 factorial design with added center reference conditions, as shown at the top of Figure 4.1.

On the worksheets the reference condition is shown at the center of the design, and in the example used for illustration the reference condition was actually so located. However, the same worksheet and identical calculations would be made irrespective of the location of an *additional* reference point.

Table 4.1. Individual batch yields in four EVOP cycles [1]

Conditions:	0	1	2	3	4
Cycle 1	63.7	62.8	63.2	67.2	60.5
Cycle 2	62.1	65.8	65.5	67.6	61.3
Cycle 3	59.6	62.1	62.0	65.3	64.1
Cycle 4	63.5	62.8	67.9	62.6	61.7

[1] Prior estimate of $\sigma = 1.8$.

We here follow only a single response—the process yield. Often several responses—yield, cost, level of impurity, etc.—would need to be studied, in which case calculations would be carried through for each response on separate worksheets.

Figure 4.1 shows the entries on the information board as they appeared at the end of the fourth cycle. Worksheet calculations were, of course, made as each cycle was completed and the information board was appropriately up-dated.

We will carry through the worksheet calculations for the first three of these cycles. For practice the reader may wish to complete the worksheet for the fourth cycle himself and check that his results agree with those shown on the information board in Figure 4.1.

The yield results obtained for the first four cycles and the prior estimate $\sigma = 1.8$ obtained from earlier phases are displayed in Table 4.1.*

* The calculation procedures, and a slightly modified form of the example, were given by Box and Hunter (1959).

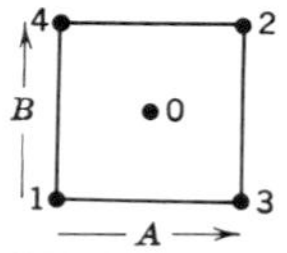

2^2 *Factorial with added reference condition*

CYCLE n = 1

Response YIELD

Project D/26-3
Phase 6
Date 2 Feb 58

Calculation of Averages					
Operating conditions	(0)	(1)	(2)	(3)	(4)
(i) Previous cycle sum					
(ii) Previous cycle average					
(iii) New observations	63.7	62.8	63.2	67.2	60.5
(iv) Differences (ii) less (iii)					
(v) New sums	63.7	62.8	63.2	67.2	60.5
(vi) New averages: $\bar{y}_i$	63.7	62.8	63.2	67.2	60.5

Calculation of Standard Deviation

Prior estimate of σ = 1.8*

Previous sum s =

Previous average s =

New s = range $\times f_{5,n}$ =

Range =

New sum s =

New average $s = \dfrac{\text{new sum } s}{n-1}$ =

Calculation of Effects

Phase mean $= \frac{1}{5}(\bar{y}_0 + \bar{y}_1 + \bar{y}_2 + \bar{y}_3 + \bar{y}_4) =$ 63.5

(A) TIME effect $= \frac{1}{2}(\bar{y}_2 + \bar{y}_3 - \bar{y}_1 - \bar{y}_4)$ = 3.6

(B) TEMP effect $= \frac{1}{2}(\bar{y}_2 + \bar{y}_4 - \bar{y}_1 - \bar{y}_3)$ = −3.2

(AB) TIME X TEMP effect $= \frac{1}{2}(\bar{y}_1 + \bar{y}_2 - \bar{y}_3 - \bar{y}_4)$ = −0.8

Change in mean effect = phase mean $- \bar{y}_0$ = −0.2

Calculation of 2 S.E. Limits

For new average s: $\pm \dfrac{2}{\sqrt{n}} s$ = ±3.6*

For new effects: $\pm \dfrac{2}{\sqrt{n}} s$ = ±3.6*

For change in mean: $\pm \dfrac{1.79}{\sqrt{n}} s$ = ±3.2*

Table of Multiplying Factors

n	1	2	3	4	5	6	7	8	9	10
$f_{5,n}$		0.30	0.35	0.37	0.38	0.39	0.40	0.40	0.40	0.41
$1/n$	1.00	0.50	0.33	0.25	0.20	0.17	0.14	0.12	0.11	0.10
$1/(n-1)$		1.00	0.50	0.33	0.25	0.20	0.17	0.14	0.12	0.11
$2/\sqrt{n}$	2.00	1.41	1.15	1.00	0.89	0.82	0.76	0.71	0.67	0.63
$1.79/\sqrt{n}$	1.79	1.26	1.03	0.89	0.80	0.73	0.68	0.63	0.60	0.57

Figure 4.2. *Two-variable EVOP program calculation worksheet. Calculations for cycle 1.*

2^2 Factorial with added reference condition

B, 0, 1, 3, A →

CYCLE $n = 2$

Response YIELD

Project D/26-3
Phase 6
Date 4 Feb 58

Calculation of Averages

Operating conditions	(0)	(1)	(2)	(3)	(4)
(i) Previous cycle sum	63.7	62.8	63.2	67.2	60.5
(ii) Previous cycle average	63.7	62.8	63.2	67.2	60.5
(iii) New observations	62.1	65.8	65.5	67.6	61.3
(iv) Differences (ii) less (iii)	1.6	-3.0	-2.3	-0.4	-0.8
(v) New sums	125.8	128.6	128.7	134.8	121.8
(vi) New averages: $\bar{y}_i$	62.9	64.3	64.4	67.4	60.9

Calculation of Standard Deviation

Prior estimate of σ	=	1.8*
Previous sum s	=	
Previous average s	=	
New s = range $\times f_{5,n}$	=	1.38
Range	=	4.6
New sum s	=	1.38
New average $s = \dfrac{\text{new sum } s}{n-1}$	=	1.38

Calculation of Effects

Phase mean $= \frac{1}{5}(\bar{y}_0 + \bar{y}_1 + \bar{y}_2 + \bar{y}_3 + \bar{y}_4) =$ 64.0

(A) TIME effect $= \frac{1}{2}(\bar{y}_2 + \bar{y}_3 - \bar{y}_1 - \bar{y}_4)$ = 3.3

(B) TEMP effect $= \frac{1}{2}(\bar{y}_2 + \bar{y}_4 - \bar{y}_1 - \bar{y}_3)$ = -3.2

(AB) TIME X TEMP effect $= \frac{1}{2}(\bar{y}_1 + \bar{y}_2 - \bar{y}_3 - \bar{y}_4)$ = 0.2

Change in mean effect = phase mean $- \bar{y}_0 =$ 1.1

Calculation of 2 S.E. Limits

For new average s: $\pm \dfrac{2}{\sqrt{n}} s$ = ±2.5*

For new effects: $\pm \dfrac{2}{\sqrt{n}} s$ = ±2.5*

For change in mean: $\pm \dfrac{1.79}{\sqrt{n}} s$ = ±2.2*

Table of Multiplying Factors

n	1	2	3	4	5	6	7	8	9	10
$f_{5,n}$		0.30	0.35	0.37	0.38	0.39	0.40	0.40	0.40	0.41
$1/n$	1.00	0.50	0.33	0.25	0.20	0.17	0.14	0.12	0.11	0.10
$1/(n-1)$		1.00	0.50	0.33	0.25	0.20	0.17	0.14	0.12	0.11
$2/\sqrt{n}$	2.00	1.41	1.15	1.00	0.89	0.82	0.76	0.71	0.67	0.63
$1.79/\sqrt{n}$	1.79	1.26	1.03	0.89	0.80	0.73	0.68	0.63	0.60	0.57

Figure 4.3. *Two-variable EVOP program calculation worksheet. Calculations for cycle 2.*

2^2 *Factorial with added reference condition*

CYCLE $n = 3$

Response YIELD

Project D/26-3
Phase 6
Date 6 Feb 58

Calculation of Averages					
Operating conditions	(0)	(1)	(2)	(3)	(4)
(i) Previous cycle sum	125.8	128.6	128.7	134.8	121.8
(ii) Previous cycle average	62.9	64.3	64.4	67.4	60.9
(iii) New observations	59.6	62.1	62.0	65.3	64.1
(iv) Differences (ii) less (iii)	3.3	2.2	2.4	2.1	−3.2
(v) New sums	185.4	190.7	190.7	200.1	185.9
(vi) New averages: $\bar{y}_i$	61.8	63.6	63.6	66.7	62.0

Calculation of Standard Deviation

Prior estimate of σ	=	1.8
Previous sum s	=	1.38
Previous average s	=	1.38
New s = range $\times f_{5,n}$	=	2.28
Range	=	6.5
New sum s	=	3.66
New average $s = \dfrac{\text{new sum } s}{n-1}$	=	1.83

Calculation of Effects

Phase mean $= \frac{1}{5}(\bar{y}_0 + \bar{y}_1 + \bar{y}_2 + \bar{y}_3 + \bar{y}_4) =$ 63.5

(A) TIME effect $= \frac{1}{2}(\bar{y}_2 + \bar{y}_3 - \bar{y}_1 - \bar{y}_4)$ = 2.4

(B) TEMP effect $= \frac{1}{2}(\bar{y}_2 + \bar{y}_4 - \bar{y}_1 - \bar{y}_3)$ = −2.4

(AB) TIME × TEMP effect $= \frac{1}{2}(\bar{y}_1 + \bar{y}_2 - \bar{y}_3 - \bar{y}_4)$ = −0.8

Change in mean effect = phase mean $- \bar{y}_0 =$ 1.7

Calculation of 2 S.E. Limits

For new average s: $\pm \frac{2}{\sqrt{n}} s$ = ±2.1

For new effects: $\pm \frac{2}{\sqrt{n}} s$ = ±2.1

For change in mean: $\pm \frac{1.79}{\sqrt{n}} s$ = ±1.9

Table of Multiplying Factors

n	1	2	3	4	5	6	7	8	9	10
$f_{5,n}$		0.30	0.35	0.37	0.38	0.39	0.40	0.40	0.40	0.41
$1/n$	1.00	0.50	0.33	0.25	0.20	0.17	0.14	0.12	0.11	0.10
$1/(n-1)$		1.00	0.50	0.33	0.25	0.20	0.17	0.14	0.12	0.11
$2/\sqrt{n}$	2.00	1.41	1.15	1.00	0.89	0.82	0.76	0.71	0.67	0.63
$1.79/\sqrt{n}$	1.79	1.26	1.03	0.89	0.80	0.73	0.68	0.63	0.60	0.57

Figure 4.4. *Two-variable EVOP program calculation worksheet. Calculations for cycle 3.*

Details of the Worksheet Calculations

We now examine the completed worksheet after the first cycle shown in Figure 4.2. Since no previous observations are available, lines (i), (ii), and (iv) on this worksheet are left blank. The "New observations" from cycle 1 for each set of operating conditions are entered in line (iii). These same numbers are also written in lines (v) and (vi) because, with only a single observation available at each position, the "New sums" and "New averages" are the observations themselves. The standard deviation σ cannot be estimated from the data of a single cycle, so the entries under "Calculation of Standard Deviations" are left blank except that the prior estimate of the standard deviation ($\sigma = 1.8$) is inserted in the first line.

The entries under "Calculations of Effects" can now be made using the numbers in line (vi) which, for this first cycle, are the observations themselves. The entries in the "Calculation of 2 S. E. Limits" are obtained by substituting the prior estimate of $\sigma = 1.8$ for s. (The asterisks on the worksheet indicate use of this prior estimate.) The constants $2/\sqrt{n}$ and $1.79/\sqrt{n}$, which are needed for these calculations, are given in a table at the bottom left corner of the worksheet. (See Table VI also.)

The averages, effects, and 2 S. E. limits may now be transferred to the information board to exhibit the situation as it stands at the end of cycle 1.

As soon as the cycle 2 data is available, we can form a first estimate of σ based on data from the present phase. The steps in filling in the worksheet for the cycle 2 data are illustrated in Figure 4.3. Calculations for all subsequent cycles follows the same pattern.

STEP 1. The data from lines (v) and (vi) on the calculation worksheet of the previous cycle are copied in lines (i) and (ii) of the worksheet for the present cycle.

STEP 2. The observations recorded during the present cycle are entered in line (iii).

STEP 3. Each entry in line (iii) is subtracted from the corresponding entry in line (ii) and the result with appropriate sign attached is entered in line (iv). These differences indicate how greatly each current result differs from the average of previous experience.

STEP 4. The new sum obtained by adding the entry in line (i) to that in line (iii) is entered in line (v).

STEP 5. The new average in line (vi) is obtained by multiplying the entry in line (v) by the multiplying factor $1/n$.

STEP 6. The "Calculation of Effects" is performed in the designated space using the "New averages" of line (vi).

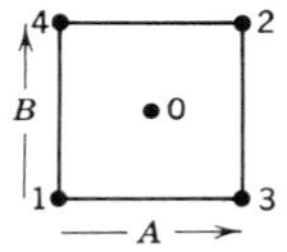

B ↑ 4 • 2, • 0, 1 • 3, A →

2^2 *Factorial with added reference condition*

CYCLE $n = 4$

Response YIELD

Project D/26-3
Phase 6
Date 8 Feb 58

Calculation of Averages

Operating conditions	(0)	(1)	(2)	(3)	(4)
(i) Previous cycle sum					
(ii) Previous cycle average					
(iii) New observations	63.5	62.8	67.9	62.8	61.7
(iv) Differences (ii) less (iii)					
(v) New sums					
(vi) New averages: $\bar{y}_i$					

Calculation of Standard Deviation

Prior estimate of σ = 1.8

Previous sum s =

Previous average s =

New s = range $\times f_{5,n}$ =

Range =

New sum s =

New average $s = \dfrac{\text{new sum } s}{n-1}$ =

Calculation of Effects

Phase mean $= \frac{1}{5}(\bar{y}_0 + \bar{y}_1 + \bar{y}_2 + \bar{y}_3 + \bar{y}_4) =$

(A) TIME effect $= \frac{1}{2}(\bar{y}_2 + \bar{y}_3 - \bar{y}_1 - \bar{y}_4) =$

(B) TEMP effect $= \frac{1}{2}(\bar{y}_2 + \bar{y}_4 - \bar{y}_1 - \bar{y}_3) =$

(AB) TIME X TEMP effect $= \frac{1}{2}(\bar{y}_1 + \bar{y}_2 - \bar{y}_3 - \bar{y}_4) =$

Change in mean effect = phase mean $- \bar{y}_0 =$

Calculation of 2 S.E. Limits

For new average s: $\pm \dfrac{2}{\sqrt{n}} s =$

For new effects: $\pm \dfrac{2}{\sqrt{n}} s =$

For change in mean: $\pm \dfrac{1.79}{\sqrt{n}} s =$

Table of Multiplying Factors

n	1	2	3	4	5	6	7	8	9	10
$f_{5,n}$		0.30	0.35	0.37	0.38	0.39	0.40	0.40	0.40	0.41
$1/n$	1.00	0.50	0.33	0.25	0.20	0.17	0.14	0.12	0.11	0.10
$1/(n-1)$		1.00	0.50	0.33	0.25	0.20	0.17	0.14	0.12	0.11
$2/\sqrt{n}$	2.00	1.41	1.15	1.00	0.89	0.82	0.76	0.71	0.67	0.63
$1.79/\sqrt{n}$	1.79	1.26	1.03	0.89	0.80	0.73	0.68	0.63	0.60	0.57

Figure 4.5. *Two-variable EVOP program calculation worksheet. Calculations for cycle 4.*

The "Calculation of the Standard Deviation" is carried out in the designated space on the right-hand side of the calculation worksheet as follows:

STEP 7. The largest and smallest differences recorded in line (iv) are underlined. The difference between these underlined values is the range (4.6 for cycle 2), and this is entered at the right-hand end of line (iv).

STEP 8. The range in line (iv) is multiplied by the factor $f_{5,n}$ to obtain the estimate s of σ contributed by this cycle. This estimate is entered in line (iii) as "New s."

A table of $f_{5,n}$ for n as large as 10 cycles is given in the bottom left-hand corner of the worksheet; also see Table V, page 222. A justification of the procedure is given in Appendix 1. At the end of cycle 2 the appropriate constant is $f_{5,2} = 0.30$, so that at this stage "New s" $= 4.6 \times 0.30 = 1.38$, and this value is entered in line (iii). Since at cycle 2 no previous estimate of the standard deviation is available from the data, the items "Previous sum s" and "Previous average s" are blank and the entries for "New sum s" and "New average s" are identical with that in line (iii).

This estimate of the standard deviation after the second cycle is not very reliable and is not actually used at this stage. It is, in fact, recorded and combined with a similar estimate from the next phase before being used to calculate 2 S. E. limits. The limits for the cycle $n = 2$ worksheet (Figure 4.3) are again obtained from the prior estimate of $\sigma = 1.8$, as indicated by the asterisks.

The up-dated averages, effects, and 2 S. E. limits showing the situations at the end of the second cycle and are now transferred to the information board.

At the end of the third cycle all the entries in the calculation worksheet can be made. The steps can be followed using the cycle $n = 3$ worksheet (Figure 4.4). The additional steps are as follows:

STEP 9. The value "New s" in line (iii) is added to the "Previous sum s" in line (i) and the result is recorded on line (v) as "New sum s." This quantity is then divided by $n - 1$ to give a "New average s" in line (vi).

STEP 10. The 2 S. E. limits for the averages and the effects using this esimate are obtained by direct substitution of "New average s" for s in the equations in the lower right portion of the calculation worksheet. The averages, effects and their error limits are now transferred to the information board to show the situation at the end of cycle 3.

The reader may wish to complete the worksheet for cycle 4 shown in

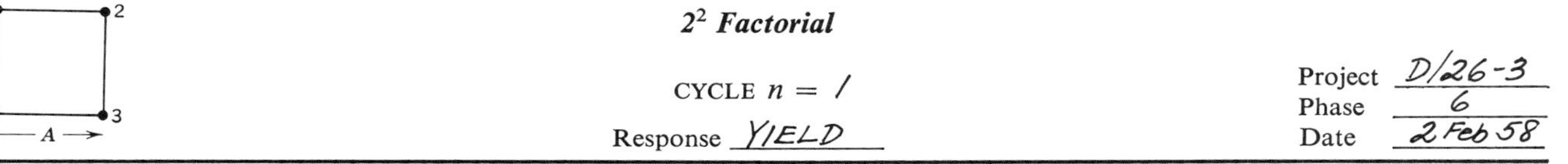

2^2 *Factorial*

CYCLE n = 1

Response YIELD

Project D/26-3
Phase 6
Date 2 Feb 58

Calculation of Averages

Operating conditions	(1)	(2)	(3)	(4)
(i) Previous cycle sum				
(ii) Previous cycle average				
(iii) New observations	62.8	63.2	67.2	60.5
(iv) Differences (ii) less (iii)				
(v) New sums	62.8	63.2	67.2	60.5
(vi) New averages: $\bar{y}_i$	62.8	63.2	67.2	60.5

Calculation of Standard Deviation

Prior estimate of σ	=	1.8*
Previous sum s	=	
Previous average s	=	
New s = range $\times f_{4,n}$	=	
Range	=	
New sum s	=	
New average $s = \frac{\text{new sum } s}{n-1}$	=	

Calculation of Effects Using Yates' Algorithm

	$y-60$	(i)	(ii)	Divisor	Effect	
(1)	2.8	10.0	13.7	4	63.4	Mean = phase mean
(3)	7.2	3.7	7.1	2	3.6	A
(4)	0.5	4.4	−6.3	2	−3.2	B
(2)	3.2	2.7	−1.7	2	−0.8	AB
S of S check	70.17		280.68	63.4−63.2=0.2		Change in mean

Calculation of 2 S.E. Limits

For new average s:	$\pm \frac{2}{\sqrt{n}} s$ =	±3.6*
For new effects:	$\pm \frac{2}{\sqrt{n}} s$ =	±3.6*
For change in mean:	$\pm \frac{1.73}{\sqrt{n}} s$ =	±3.1*

Table of Multiplying Factors

n	1	2	3	4	5	6	7	8	9	10
$f_{4,n}$		0.34	0.40	0.42	0.43	0.44	0.45	0.45	0.46	0.46
$1/n$	1.00	0.50	0.33	0.25	0.20	0.17	0.14	0.12	0.11	0.10
$1/(n-1)$		1.00	0.50	0.33	0.25	0.20	0.17	0.14	0.12	0.11
$2/\sqrt{n}$	2.00	1.41	1.15	1.00	0.89	0.82	0.76	0.71	0.67	0.63
$1.73/\sqrt{n}$	1.73	1.22	1.00	0.87	0.77	0.71	0.65	0.61	0.58	0.55

Figure 4.6. *Two-variable EVOP program calculation worksheet when there are no additional reference conditions. Calculations for cycle 1.*

Figure 4.5 and to check the values which he would transfer to the information board with those shown in Figure 4.1.

Conclusions from Figure 4.1

As we have already mentioned, it would normally be true that a number of responses—and not only yield—would be observed and considered before changing the operating conditions to new levels. It would be useless, for example, to obtain higher yields if the resulting product contained a high level of a particular impurity and was unsaleable. Considering yield alone *for this example,* however, we form the conclusion from Figure 4.1 that an increase in time would be advantageous. There is an indication (suggested also in earlier cycles) of the desirability of using a lower temperature but this would need to be investigated more fully by additional cycles or in subsequent phases.

4.3. WORKSHEETS FOR A 2^2 FACTORIAL WITHOUT ADDITIONAL REFERENCE CONDITIONS

As we have previously noted, schemes are often run in which either no reference condition is included or, alternatively, the reference condition is one of the factorial runs. An appropriate form of worksheet to cover such situations is shown in Figure 4.6. For illustration, the previous yield data have been used again with center conditions omitted, and it has been supposed that condition 2 is a reference condition. The calculations proceed essentially as before. The only differences are:

1. The factors $f_{4,n}$ are those appropriate for four sets of conditions rather than five. Appropriate values of $f_{4,n}$ are shown in the table at the bottom left corner of the worksheet.

2. As always, the change in mean effect is given by (average over all conditions) − (average at reference conditions). In the particular example this would correspond to the contrast $(1/4)(\bar{y}_1 + \bar{y}_3 + \bar{y}_4 - 3\bar{y}_2)$. The 2 S. E. limits are given by [change in mean $\pm(1.73/\sqrt{n})s$]. The constant $1.73/\sqrt{n}$ is given in the table at the bottom left-hand corner of the worksheet (Figure 4.6). (See Table VI also.)

When the reference conditions form one of the points of the factorial design, the "Change in mean" effect will not be independent of the other calculated effects. It will, in fact, be an explicit linear function of these quantities. This does not mean, of course, that it is not worth calculating. It allows a different aspect of the data to be appreciated and so it is of value. The same worksheet can, of course, be used when none of the factorial points are reference conditions. In this case we merely omit the calculation of the change in mean and its 2 S. E. limits.

Calculation of Effects Using Yates' Algorithm

For the 2^2 factorial the direct calculation of the effect is so simple that it is doubtful whether the use of Yates' algorithm is justified. Some prefer to use this algorithm, however, and in Figure 4.6 the opportunity has been taken to show an appropriate arrangement on the worksheet. Of course, if desired, a modified worksheet using Yates' algorithm can also be used with added reference conditions.

In Figure 4.6 it will be noted that for simplicity the Yates' calculations have been carried through with 60 subtracted from all the values.* This quantity is then added to the final mean. The sum of squares check is applied as before. In the sample calculation $280.68 = 4 \times 70.17$, which verifies the correctness of the computation.

* It is easy to verify that if *any* constant K (positive or negative) is subtracted from all the average response values used in a Yates' calculation all the factorial effects are unchanged and the over-all mean is reduced by K. By a suitable choice of K (e.g., 60 in Figure 4.6) we can thus slightly simplify the Yates' calculation arithmetic.

CHAPTER 5

Worksheets for Three-variable *EVOP* Programs

5.1. INTRODUCTION

We now discuss the construction and use of worksheets for three-variable EVOP programs employing the 2^3 factorial design. To eliminate removable extraneous variation while running such programs, the first half of any given cycle and the second half of that cycle are scheduled so as to correspond to two blocks of the 2^3 design in the manner described in Chapter 3.

The worksheets for three-variable programs are very similar to those for two-variable programs, but we now employ *two* worksheets for each cycle: one for the first block and one for the second. After the first cycle a new estimate of the error variance σ will be available *after each new block* (i.e., after each half-cycle). New estimates of the effects will be available only after each complete cycle. As before, the worksheets and the calculations will differ slightly depending on whether or not a reference condition is included in the design and, if it is included, on whether this condition corresponds to one of the factorial runs or is an additional run.

In this chapter we first illustrate in some detail the use of worksheets appropriate when the reference condition is not one of the factorial runs. In our model scheme the reference condition is run once in any *block* (i.e., twice per cycle). Also, in the particular example used, this reference condition happens to be at the center of the factorial design. The same worksheets and calculations could, of course, be used whatever the location of this additional reference condition. Later in the chapter we show the slight modification which is necessary if there is no additional reference condition.

5.2. WORKSHEETS FOR A 2^3 FACTORIAL ARRANGED IN TWO BLOCKS WITH A REFERENCE RUN IN EACH BLOCK

The following illustration of the use of the worksheets is based on data * coming from the third phase of an EVOP program applied to the manufacture of an antibiotic. In this phase, residence time, reaction temperature and pH were being investigated. The response followed was yield, which was extremely variable. Figure 5.1 shows the layout of the design. The numbering refers to the sequence in which the runs were made. Runs 0, 1, 2, 3, 4, shown by filled dots, comprise block I, and runs Ø, 5, 6, 7, 8, shown by open dots, comprise block II. The data in Table 5.1 are the yield results for the first three cycles.

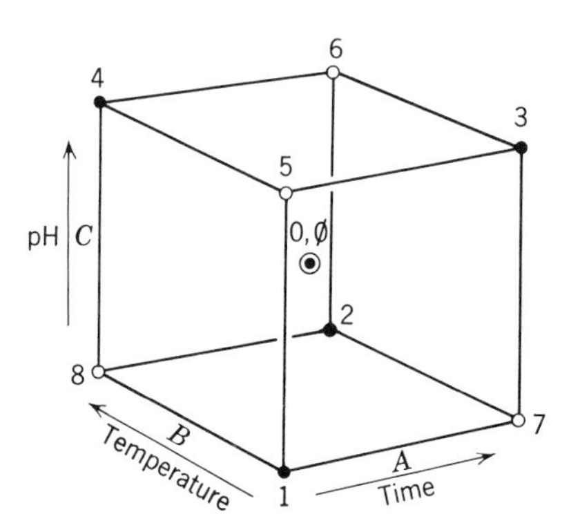

Figure 5.1. A 2^3 design with center points, used in a three-variable EVOP scheme.

The calculations included on the worksheets parallel very closely those used in the two-variable scheme, and the entries on the sheets (see Figures 5.2 through 5.7) will be largely self-explanatory. It should be noticed, however, that, since there are 10 sets of operating conditions (operating conditions 0 and Ø at the center of the cube are, in fact, identical but are treated as separate since they are performed in different blocks), we must keep tally on 10 cumulative averages so that, for example, the preceding sum for condition 0 will be found on the preceding worksheet for block I and not on the worksheet for block II.

* The calculation procedures, and a slightly modified form of the example, were given by Box and Hunter (1959).

Table 5.1. Individual run yields for three EVOP cycles [1]

	Block I					Block II				
Conditions	0	1	2	3	4	Ø	5	6	7	8
Cycle 1	78	82	63	81	88	85	79	75	78	67
Cycle 2	82	75	79	96	77	69	77	80	70	84
Cycle 3	65	82	68	85	79	87	96	66	82	72

[1] Prior estimate of $\sigma = 8.0$.

From cycle $n = 2$ onward an estimate of the standard deviation is computed from the data of each block and is combined with estimates from previous blocks. As before, a prior estimate of σ is used until (at the end of cycle 2) two current estimates are available for averaging. (As before, asterisks indicate use of the prior estimate in the first cycle.)

With the completion of each cycle (each pair of blocks) the effects may be recomputed using Yates' algorithm. The worksheet for block II provides a suitable format for this calculation.* The various multipliers $f_{5,n}$, $1/n$, $1/[2(n-1)]$, $2/\sqrt{n}$, $1.41/\sqrt{n}$, and $1.26/\sqrt{n}$, which are needed at various stages are listed for $n = 1, 2, \ldots, 6$ on the block I sheets. Thus, for example, the quantity "Average s" is obtained by multiplying "New sum s" by the multiplier $1/[2(n-1)]$ which is listed in the third row of the table of multiplying factors. (See Table VI also.)

At the extreme right-hand bottom corner of the block II worksheet, the calculations yielding the phase mean, the reference mean, and the change in mean are set out. Again the arrangement is self-explanatory.

Summary of Calculations

To ensure that calculations are correctly performed, it is a good idea to have a complete set of instructions printed on the back of each worksheet to act as a reminder and a reference. A suitable layout for the back of the block II worksheet is shown in Figure 5.8.

The information board as it would appear at the end of Cycle 3, after transfer of the various calculated quantities from the appropriate worksheet, is shown in Figure 5.9. The reader might like to consider what tentative conclusions he could draw from this display. (See page 129.)

* On the block II worksheets, a constant has been subtracted from all averages for the Yates' and other calculations; see the footnote on page 118.

2^3 *Factorial in two blocks with reference run in each block*

BLOCK I

CYCLE n = 1

Response YIELD

Project WY
Phase 3
Date 29 JAN

Calculation of Averages						Calculation of Standard Deviation
Operating conditions	(0)	(1)	(2)	(3)	(4)	
(i) Previous sum for block I						Previous sum s (all blocks) =
(ii) Previous average for block I						
(iii) New observations for block I	78	82	63	81	88	New s = range $\times f_{5,n}$ =
(iv) Differences (ii) less (iii)						Range =
(v) New sums for block I	78	82	63	81	88	New sum s (all blocks) =
(vi) New averages for block I	78	82	63	81	88	

Table of Multiplying Factors

n	1	2	3	4	5	6
$f_{5,n}$		0.30	0.35	0.37	0.38	0.39
$1/n$	1	0.50	0.33	0.25	0.20	0.17
$1/[2(n-1)]$		0.50	0.25	0.17	0.12	0.10
$2/\sqrt{n}$	2.00	1.41	1.15	1.00	0.89	0.82
$1.41/\sqrt{n}$	1.41	1.00	0.82	0.71	0.63	0.58
$1.26/\sqrt{n}$	1.26	0.89	0.73	0.63	0.57	0.52

Figure 5.2. *Three-variable EVOP program calculation worksheet. Block I calculations for cycle 1.*

2^3 *Factorial in two blocks with reference run in each block*

BLOCK II

CYCLE n = 1

Response YIELD

Project WY
Phase 3
Date 31 JAN

Calculation of Averages					
Operating conditions	(ø)	(5)	(6)	(7)	(8)
(i) Previous sum for block II					
(ii) Previous average for block II					
(iii) New observations for block II	85	79	75	78	67
(iv) Differences (ii) less (iii)					
(v) New sums for block II	85	79	75	78	67
(vi) New averages for block II	85	79	75	78	67

Calculation of Standard Deviation		
Prior estimate of σ	=	8*
Previous sum s (all blocks)	=	
New s = range $\times f_{5,n}$	=	
Range	=	
New sum s (all blocks)	=	
Average $s = \{1/[2(n-1)]\} \times$ new sum s	=	

Calculation of Effects Using Yates' Algorithm

	$\bar{y}-60$	(i)	(ii)	(iii)	Multiplier	Effect	
(1)	22	40	50	133	0.125	16.6	Factorial mean − 60
(7)	18	10	83	−19	0.25	-4.75	A
(8)	7	40	−8	−27	0.25	-6.75	B
(2)	3	43	−11	−15	0.25	-3.75	AB
(5)	19	−4	−30	33	0.25	8.25	C
(3)	21	−4	3	−3	0.25	-0.75	AC
(4)	28	2	0	33	0.25	8.25	BC
(6)	15	−13	15	−15	0.25	-3.75	ABC + blocks
S of S check	2,677			21,416			

Calculation of 2 S.E. Limits

For new averages: $\pm \frac{2}{\sqrt{n}} s =$ ± 16*

For new effects: $\pm \frac{1.41}{\sqrt{n}} s =$ ± 11.3*

For change in mean: $\pm \frac{1.26}{\sqrt{n}} s =$ ± 10.1*

	$\bar{y}-60$	Sum	Multiplier	Mean −60		
Reference Conditions (0) (ø)	18 25	43	0.5	21.5	Reference mean	81.5
Factorial conditions		133	0.125	16.6	Factorial mean	76.6
All conditions		176	0.1	17.6	Phase mean	77.6
Change in mean =		77.6	-81.5	= -3.9		

Figure 5.3. *Three-variable EVOP program calculation worksheet. Block II calculations for cycle 1.*

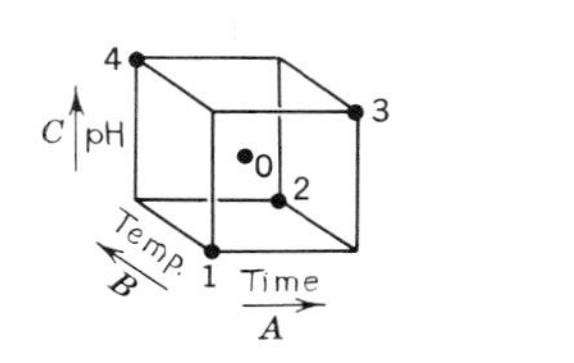

2^3 *Factorial in two blocks with reference run in each block*

BLOCK I

CYCLE $n = 2$

Response YIELD

Project WY
Phase 3
Date 2 FEB

Calculation of Averages						Calculation of Standard Deviation
Operating conditions	(0)	(1)	(2)	(3)	(4)	
(i) Previous sum for block I	78	82	63	81	88	Previous sum s (all blocks) =
(ii) Previous average for block I	78	82	63	81	88	
(iii) New observations for block I	82	75	79	96	77	New s = range $\times f_{5,n}$ = 8.1
(iv) Differences (ii) less (iii)	−4	7	−16	−15	11	Range = 27
(v) New sums for block I	160	157	142	177	165	New sum s (all blocks) = 8.1
(vi) New averages for block I	80.0	78.5	71.0	88.5	82.5	

Table of Multiplying Factors

n	1	2	3	4	5	6
$f_{5,n}$		0.30	0.35	0.37	0.38	0.39
$1/n$	1	0.50	0.33	0.25	0.20	0.17
$1/[2(n-1)]$		0.50	0.25	0.17	0.12	0.10
$2/\sqrt{n}$	2.00	1.41	1.15	1.00	0.89	0.82
$1.41/\sqrt{n}$	1.41	1.00	0.82	0.71	0.63	0.58
$1.26/\sqrt{n}$	1.26	0.89	0.73	0.63	0.57	0.52

Figure 5.4. *Three-variable EVOP program calculation worksheet. Block I calculations for cycle 2.*

2^3 Factorial in two blocks with reference run in each block

BLOCK II

CYCLE $n = 2$

Response YIELD

Project WY
Phase 3
Date 4 FEB

Calculation of Averages

Operating conditions	(ø)	(5)	(6)	(7)	(8)
(i) Previous sum for block II	85	79	75	78	67
(ii) Previous average for block II	85	79	75	78	67
(iii) New observations for block II	69	77	80	70	84
(iv) Differences (ii) less (iii)	16	2	−5	8	−17
(v) New sums for block II	154	156	155	148	151
(vi) New averages for block II	77.0	78.0	77.5	74.0	75.5

Calculation of Standard Deviation

Prior estimate of σ	=	8
Previous sum s (all blocks)	=	8.1
New s = range $\times f_{5,n}$	=	9.9
Range	=	33
New sum s (all blocks)	=	18.0
Average $s = \{1/[2(n-1)]\} \times$ new sum s	=	9.0

Calculation of Effects Using Yates' Algorithm

	$\bar{y} - 70$	(i)	(ii)	(iii)	Multiplier	Effect	
(1)	8.5	12.5	19.0	65.5	0.125	8.19	Factorial mean − 70
(7)	4.0	6.5	46.5	−3.5	0.25	−0.88	A
(8)	5.5	26.5	−9.0	−12.5	0.25	−3.12	B
(2)	1.0	20.0	5.5	−15.5	0.25	−3.88	AB
(5)	8.0	−4.5	−6.0	27.5	0.25	6.88	C
(3)	18.5	−4.5	−6.5	14.5	0.25	3.62	AC
(4)	12.5	10.5	0.0	−0.5	0.25	−0.12	BC
(6)	7.5	−5.0	−15.5	−15.5	0.25	−3.88	ABC + blocks
S of S check	738.25		5906.00				

Calculation of 2 S.E. Limits

For new averages: $\pm \frac{2}{\sqrt{n}} s = \pm 12.7$

For new effects: $\pm \frac{1.41}{\sqrt{n}} s = \pm 9.0$

For change in mean: $\pm \frac{1.26}{\sqrt{n}} s = \pm 8.0$

	$\bar{y} - 70$	Sum	Multiplier	Mean − 70		
Reference Conditions (0) / (ø)	10.0 / 7.0	17.0	0.5	8.5	Reference mean	78.5
Factorial conditions		65.5	0.125	8.19	Factorial mean	78.19
All conditions		82.5	0.1	8.25	Phase mean	78.25
Change in mean =		78.25	− 78.5	= −0.25		

Figure 5.5. *Three-variable EVOP program calculation worksheet. Block II calculations for cycle 2.*

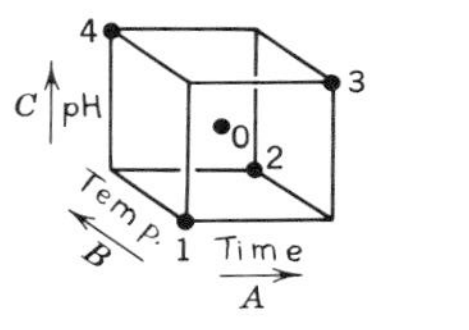

2^3 *Factorial in two blocks with reference run in each block*

BLOCK I

CYCLE $n = 3$

Response YIELD

Project WY
Phase 3
Date 7 FEB

Calculation of Averages						Calculation of Standard Deviation	
Operating conditions	(0)	(1)	(2)	(3)	(4)		
(i) Previous sum for block I	160	157	142	177	165	Previous sum s (all blocks) =	18.0
(ii) Previous average for block I	80.0	78.5	71.0	88.5	82.5		
(iii) New observations for block I	65	82	68	85	79	New s = range $\times f_{5,n}$ =	6.5
(iv) Differences (ii) less (iii)	15.0	-3.5	3.0	3.5	3.5	Range =	18.5
(v) New sums for block I	225	239	210	262	244	New sum s (all blocks) =	24.5
(vi) New averages for block I	75.0	79.3	70.0	87.3	81.3		

Table of Multiplying Factors

n	1	2	3	4	5	6
$f_{5,n}$		0.30	0.35	0.37	0.38	0.39
$1/n$	1	0.50	0.33	0.25	0.20	0.17
$1/[2(n-1)]$		0.50	0.25	0.17	0.12	0.10
$2/\sqrt{n}$	2.00	1.41	1.15	1.00	0.89	0.82
$1.41/\sqrt{n}$	1.41	1.00	0.82	0.71	0.63	0.58
$1.26/\sqrt{n}$	1.26	0.89	0.73	0.63	0.57	0.52

Figure 5.6. *Three-variable EVOP program calculation worksheet.' Block I calculations for cycle 3.*

C pH 8 Ø 7 Temp. B Time A

BLOCK II

CYCLE $n = 3$

Response YIELD

Project WY
Phase 3
Date 9 FEB

Calculation of Averages

Operating conditions	(Ø)	(5)	(6)	(7)	(8)
(i) Previous sum for block II	154	156	155	148	151
(ii) Previous average for block II	77.0	78.0	77.5	74.0	75.5
(iii) New observations for block II	87	96	66	82	72
(iv) Differences (ii) less (iii)	-10.0	-18.0	11.5	-8.0	3.5
(v) New sums for block II	241	252	221	230	223
(vi) New averages for block II	80.3	84.0	73.7	76.7	74.3

Calculation of Standard Deviation

Prior estimate of σ	=	8
Previous sum s (all blocks)	=	24.5
New s = range $\times f_{5,n}$	=	10.3
Range	=	
New sum s (all blocks)	=	34.8
Average $s = \{1/[2(n-1)]\} \times$ new sum s	=	8.7

Calculation of Effects Using Yates' Algorithm

	$\bar{y} - 70$	(i)	(ii)	(iii)	Multiplier	Effect	
(1)	9.7	16.4	20.7	67.0	0.125	8.38	Factorial mean − 70
(7)	6.7	4.3	46.3	-11.6	0.25	-2.90	A
(8)	4.3	31.3	-7.3	-28.4	0.25	-7.10	B
(2)	0.0	15.0	-4.3	-12.2	0.25	-3.05	AB
(5)	14.0	-3.0	-12.1	25.6	0.25	6.40	C
(3)	17.3	-4.3	-16.3	3.0	0.25	0.75	AC
(4)	11.3	3.3	-1.3	-4.2	0.25	-1.05	BC
(6)	3.7	-7.6	-10.9	-9.6	0.25	-2.40	ABC + blocks
S of S check	794.14			6353.12			

Calculation of 2 S.E. Limits

For new averages: $\pm \frac{2}{\sqrt{n}} s = \pm 10.0$

For new effects: $\pm \frac{1.41}{\sqrt{n}} s = \pm 7.1$

For change in mean: $\pm \frac{1.26}{\sqrt{n}} s = \pm 6.3$

	$\bar{y} - 70$	Sum	Multiplier	Mean − 70	
Reference Conditions (0) / (Ø)	5.0 / 10.3	15.3	0.5	7.65	Reference mean 77.65
Factorial conditions		67.0	0.125	8.38	Factorial mean 78.38
All conditions		82.3	0.1	8.23	Phase mean 78.23
Change in mean =		78.23	-77.65	= 0.58	

Figure 5.7. Three-variable EVOP program calculation worksheet. Block II calculations for cycle 3.

Block II Summary of Calculations

1. Enter project number, phase, date, response, and cycle number n.

CALCULATION OF AVERAGES

2. Copy lines (i) and (ii) from lines (v) and (vi) of block II worksheet for last cycle.
3. Enter new observations corresponding to conditions Ø, 5, 6, 7 and 8 in line (iii).
4. Calculate differences (ii) less (iii) and enter in (iv). Underline largest and smallest entries, taking full account of sign.
5. Add entries in lines (i) and (iii) and enter in line (v).
6. Multiply entries in line (v) by $1/n$ (see table of multiplying constants). Enter results in line (vi).

CALCULATION OF STANDARD DEVIATION

7. Enter "Previous sum s" (all blocks) from *block I worksheet* in line (i).
8. Calculate range from difference of largest and smallest entries underlined on line (iv) (left of page) and enter in line (iv) on right of page.
9. Multiply range in line (iv) by $f_{5,n}$ (see table of multiplying constants) to obtain "New s." Enter on line (iii).
10. Add entries in lines (iii) and (i) to obtain "New sum s" all blocks. Multiply by constant $1/[2(n-1)]$ (see table of multiplying constants) to obtain "Average s." Enter in line (vi).

CALCULATION OF EFFECTS USING YATES' ALGORITHM

11. After subtracting suitable constant (record at top of column) enter "New averages" in order shown. Sum squares.

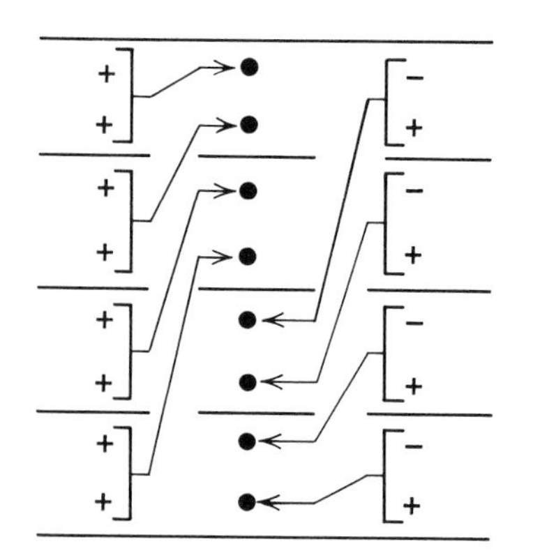

12. Perform Yates' additions and subtractions indicated in the diagram to obtain column (i). Repeat these operations with elements of column (i) to obtain column (ii) and with elements of column (ii) to obtain column (iii).
13. Check by computing sum of squares of column (iii). This should total *exactly* 8 times the sum of squares calculated at step (11).
14. Multiply entries in column (iii) by multipliers to obtain "Effects."
15. Enter observations for reference conditions 0 and Ø on right of worksheet (subtract suitable constant).
16. Compute "Reference mean," "Factorial mean," and "Phase mean" as shown (add back constant).
17. Obtain "Change in mean" from phase mean minus reference mean.

CALCULATION OF 2 S.E. LIMITS

18. Compute factors for "2 S.E. Limits" using multipliers in table.

Figure 5.8. *Example of summarized calculation instructions printed on reverse of block II worksheet.*

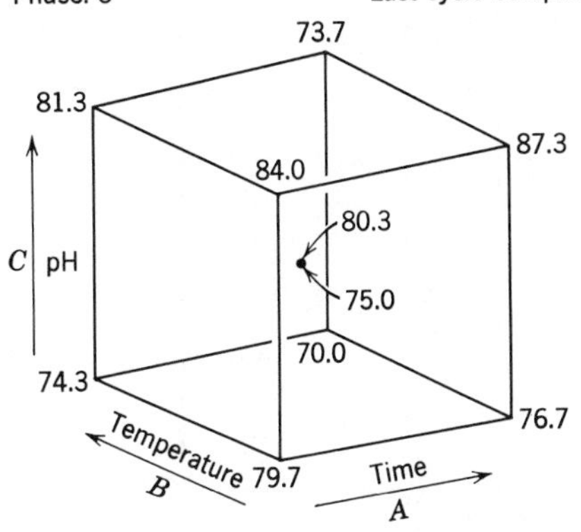

2 S. E. Limits for individual averages: ± 10.0

		Yield	Other Responses
Requirement:		Maximize	...
	Phase mean	78.2	...
Effects with 2 S.E. limits	Time	-2.90 ± 7.1	
	Temp.	-7.10 ± 7.1	
	pH	6.40 ± 7.1	
	Time $\times$ temp.	-3.05 ± 7.1	
	Time $\times$ pH	0.75 ± 7.1	
	Temp. $\times$ pH	-1.05 ± 7.1	
	Change in mean	0.58 ± 6.3	...

Figure 5.9. *Part of a three-variable EVOP information board after three cycles.*

Checks

A wise philosophy is to treat all calculations as probably wrong until proved otherwise. The checks applied may be formal ones such as the sum of squares check in the Yates calculation or informal ones. As an example of an informal check, consider the information board after three cycles shown in Figure 5.9. According to the entries shown there, a negative temperature effect and a positive pH effect are beginning to be distinguishable from error. These ought, therefore, to be evident in the averages displayed at the corners of the cube immediately above the analysis. Inspection of the appropriate values on the cube confirms this. On the other hand, if some large "effect" had been found which was not evident on inspection of the data displayed in the cube, we should suspect an error in the calculations.

5.3. WORKSHEETS FOR A BLOCKED 2^3 FACTORIAL WITHOUT ADDITIONAL REFERENCE RUNS

Worksheets for a 2^3 design without additional reference runs are a little different from the ones previously shown. A suitable format is given in

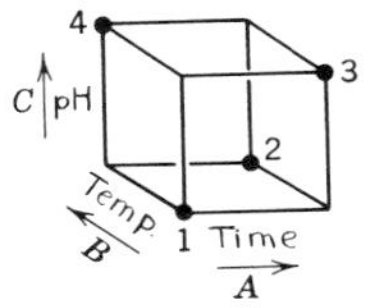

C pH
4
3
2
Temp. B
1 Time A

2^3 *Factorial in two blocks*

BLOCK I

CYCLE $n = 1$

Response YIELD

Project WY
Phase 3
Date 29 JAN

Calculation of Averages					Calculation of Standard Deviation
Operating conditions	(1)	(2)	(3)	(4)	Prior estimate of σ =
(i) Previous sum for block I					Previous sum s (all blocks) =
(ii) Previous average for block I					
(iii) New observations for block I	82	63	81	88	New s = range $\times f_{4,n}$ =
(iv) Differences (ii) less (iii)					Range =
(v) New sums for block I	82	63	81	88	New sum s (all blocks) =
(vi) New averages for block I	82	63	81	88	

Table of Multiplying Factors

n	1	2	3	4	5	6
$f_{4,n}$		0.34	0.40	0.42	0.43	0.44
$1/[2(n-1)]$		0.50	0.25	0.17	0.12	0.10
$2/\sqrt{n}$	2.00	1.41	1.15	1.00	0.89	0.82
$1.41/\sqrt{n}$	1.41	1.00	0.82	0.71	0.63	0.58

Figure 5.10. *Three-variable EVOP program calculation worksheet when there are no additional reference conditions. Block I calculations for cycle 1.*

C pH
Temp. B
Time A
5 6 7 8

2^3 *Factorial in two blocks*

BLOCK II
CYCLE n = 1
Response YIELD

Project WY
Phase 3
Date 31 JAN

Calculation of Averages				
Operating conditions	(5)	(6)	(7)	(8)
(i) Previous sum for block II				
(ii) Previous average for block II				
(iii) New observations for block II	79	75	78	67
(iv) Differences (ii) less (iii)				
(v) New sums for block II	79	75	78	67
(vi) New averages for block II	79	75	78	67

Calculation of Standard Deviation		
Prior estimate of σ	=	8*
Previous sum s (all blocks)	=	
New s = range $\times f_{4,n}$	=	
Range	=	
New sum s (all blocks)	=	
Average $s = \{1/2[(n-1)]\} \times$ new sum s	=	

Calculation of Effects Using Yates' Algorithm

	$\bar{y} - 60$	(i)	(ii)	(iii)	Multiplier	Effect	
(1)	22	40	50	133	0.125	16.6	Factorial mean − 60
(7)	18	10	83	−19	0.25	−4.75	A
(8)	7	40	−8	−27	0.25	−6.75	B
(2)	3	43	−11	−15	0.25	−3.75	AB
(5)	19	−4	−30	33	0.25	8.25	C
(3)	21	−4	3	−3	0.25	−0.75	AC
(4)	28	2	0	33	0.25	8.25	BC
(6)	15	−13	15	−15	0.25	−3.75	ABC + blocks
S of S check	2,677			21,416			

Calculation of 2 S.E. Limits

For new averages: $\pm \frac{2}{\sqrt{n}} s =$ ±16*

For new effects: $\pm \frac{1.41}{\sqrt{n}} s =$ ±11.3*

Phase mean = factorial mean = 77.6

Figure 5.11. *Three-variable EVOP program calculation worksheet when there are no additional reference conditions. Block II calculations for cycle 1.*

Figures 5.10 and 5.11 and calculations for a first cycle are shown, using the previous data. The main difference is the use of the factor $f_{4,n}$ instead of $f_{5,n}$ in estimating the standard deviation from the range. Also, no provision has been made for calculating a change in mean. This can be accommodated, but we must note the following. If one of the factorial runs is a reference condition and it is run in only one block, the over-all change in mean can be calculated as usual, but it is influenced by block-to-block variation (see Section 5.5). If one of the factorial runs is a reference condition and it is run in *every* block, the over-all change in mean is free from block to block variation, but the "extra" reference run, that is, the one that lies in the block to which it does not belong, should be ignored when the factorial effects are calculated. Although these EVOP variations are a little more complex than those preceding, they can still be handled by using the principles we describe in this book.

5.4. SIMPLICITY AND SOPHISTICATION

Evolutionary Operation is a simple technique designed to be used with elementary tools. The foregoing worksheets are such that, after some practice, an operator specially detailed to the task will be employed for only short periods, perhaps for 30 minutes a day, in keeping the information board up to date.

As we have said, in the more industrially developed countries, the availability of computers or input terminals at manufacturing locations is becoming so widespread that, in some instances, EVOP calculations are being made with the help of a computer. In circumstances where even moderately unskilled labor is expensive, it is much cheaper to use computers, when they are already available, to make calculations. Once the "bugs" have been removed from the program, the computer provides a rapid and accurate means of updating the information board.

It should not be thought, however, that computers are *necessary* to the carrying out of EVOP. One of the principal attractions of this procedure is that sophistication is not needed, and the method has been used with success even under the most primitive conditions.

5.5. BLOCK-TO-BLOCK VARIATION AND THE STANDARD ERROR OF THE PHASE MEAN

It has been pointed out that the variation between runs made close together in time is usually less than the variation between runs made further apart. For this reason greater accuracy can be obtained by arranging things so that effect contrasts are calculated entirely from comparisons made *within blocks*. We have noted, however, that in addition to *effect* contrasts the process superintendent will be interested in the *phase mean* which measures the *absolute* performance of the process during a particular phase. For each particular response the phase mean will be estimated by an average taken over all conditions for all cycles in the phase or, equivalently, by the average of all the block averages recorded during the phase.

Suppose that at the end of a particular cycle in a particular phase N runs have been made. It might at first be thought that the standard error of the phase mean will be $s/\sqrt{N}$, where s is the estimated standard deviation calculated on the worksheet in the manner we have described. This estimate would usually be inappropriate because s is an estimate only of the standard deviation *within* blocks.

The standard deviation within blocks is, of course, all that concerns us in assessing possible errors in effect contrasts which are unaffected by block-to-block differences. However, the phase mean is an absolute measure of performance estimated by an average taken *across* blocks and so clearly is influenced by variation *from block to block*. If 2 S. E. limits of the phase mean are required, these are usually obtained from an estimate of the standard deviation of block averages recorded during the current and previous phases. If the estimate is s_b and b blocks have been completed in the current phase, then the standard error of the phase mean is $s_b/\sqrt{b}$.

For illustration, Table 5.2 shows the *block averages* for the current phase 3 and for the previous two phases. Using these data, estimates of the standard deviation for block averages are obtained from each phase by multiplying the range by the appropriate multiplying factor w listed in

Table 5.2. Calculation of weighted average estimate of standard deviation of block averages

Phase	Block Averages	Range	b Number of Blocks	w_b	s_b	$b-1$ Approximate Weight
1	<u>66.0</u>, 69.9, 73.6, 71.4, 74.4, 70.9, <u>77.5</u>, 67.6,	11.5	8	0.351	4.04	7
2	74.1, <u>79.8</u>, 75.2, <u>68.1</u>	11.7	4	0.486	5.69	3
3	78.4, 76.8, <u>81.8</u>, 76.0, <u>75.8</u>, 80.6	6.0	6	0.395	2.37	5

Weighted average estimate of standard deviation of block averages $\bar{s}_b$

$$= \frac{(7 \times 4.04) + (3 \times 5.69) + (5 \times 2.37)}{7 + 3 + 5} = 3.81$$

Table 2.3. Because the number of block averages is not the same in all blocks, a weighted estimate $\bar{s}_b$ of the standard deviation of block averages is calculated. Approximate weights,* equal to the number of blocks in each cycle less one, are used. The weighted estimate $\bar{s}_b$ may be used to calculate the 2 S. E. limits for the phase 3 mean as follows.

$$\text{Standard error for phase 3 mean} = \frac{\bar{s}_b}{\sqrt{6}} = \frac{3.81}{\sqrt{6}} = 1.56.$$

$$\text{Estimated phase 3 mean with 2 S. E. limits} = 78.2 \pm 3.1.$$

In this particular example each block contained five observations, and it will be recalled that the prior estimate for σ *within blocks* was 8.0. If there were no additional variation from block to block, the standard deviation between block means would be $8.0/\sqrt{5} = 3.58$. This is to be compared with the standard deviation of the block averages of 3.81. In this particular example, therefore, the variation between block averages was only a little larger than what might have been expected from the within-block variance. In the example then, little appears to be gained by blocking, although this will not usually be true.

Precisely similar calculations may, of course, be made for a two-factor EVOP scheme. In this case the individual cycles would correspond to the blocks.

An estimate s_b can be obtained at the commencement of an EVOP program from previous data. A period should be picked where no known changes have been introduced into the process. If the proposed EVOP block size is k (usually $k = 4$ or 5), then successive groups of observations of size k are marked off, their means calculated, and the standard deviation of these means obtained in the usual way.

* A more precise method of weighting range estimates has been given by H. A. David (1951). The above method provides an adequate approximation when, as is usually the case, the number of blocks in each cycle is between 2 and 10.

CHAPTER 6

Some Aspects of the Organization of Evolutionary Operation

6.1. TRAINING PROGRAM

For EVOP to be properly effective, it must be understood, appreciated, and supported by people at various levels in a company. Thus some sort of educational and orientational program is needed so that everyone can appreciate what EVOP can do and what is involved. Since different types of understanding are required by different people according to their position in the company, we distinguish between three types of education directed toward the following:

1. Higher management.
2. Supervisory personnel; for example, plant engineers, chemists, and superintendents.
3. Plant operators.

Each of these three groups has a different role in the EVOP program and, consequently, a different approach is needed for each group.

Higher Management

A successful EVOP program needs the active support of higher management. This will almost certainly be forthcoming if the nature and advantages of EVOP are clearly pointed out. A well-organized presentation lasting, perhaps, 45 minutes should be perfectly adequate for this purpose. The following points should be stressed:

1. Evolutionary Operation provides a natural, automatic improvement of processes under a planned program with well-laid lines of communication established through the EVOP committee (see Sections 1.9 and 6.3).

It avoids static plant operation, yet it does not "mess around" with the process—all departures from static operation are preplanned and controlled.

2. Evolutionary Operation uses resources already available and involves only extra organization to start the scheme and a small amount of extra supervision by personnel already available.

3. Evolutionary Operation is good for the morale of those running it. Most intelligent operators have had, at one time or another, a desire to try some set of adjustments to the plant process. Instead of prohibiting this, we now give the operators a planned program and ask them to suggest what they feel are sensible modifications for consideration by the EVOP committee. Enterprising process superintendents will also welcome EVOP since it will enable them to demonstrate their abilities in improving the process under their control.

4. The EVOP committee provides an opportunity for the working of the process to be critically examined at regular intervals by a group of experts. These include not only those directly responsible for the running of the process but also a group of specialists whose talents are normally employed elsewhere (in research, quality control, etc.), who will periodically look at the process from a fresh viewpoint (see Section 6.3).

The general philosophy of EVOP has a natural appeal to higher management. Although they need to know, in broad outline, the way in which an EVOP program is run, it is unnecessary to go into great detail. A talk based on the first chapter of this book should be sufficient.

Supervisory Personnel

Plant engineers, chemists, process superintendents, and others who will supervise the EVOP scheme and/or be present on the EVOP committee need to have a thorough understanding of EVOP, and it is at this level that the maximum training effort has to be applied. Experience with a number of training courses that have been presented over the last 10 years suggests that at least a full two-day training course is needed to cover the necessary material, which should include all the aspects of EVOP discussed in this book. When possible, the best results are probably achieved by spreading teaching over, say, a period of two weeks, teaching two half-days each week; exercises are worked on by the students in the interim.

A particularly important aspect of the course is the simulation game discussed in Section 6.2. The game enables the students to learn by doing, not only by carrying out the calculations but also by discussing the results at the end of each cycle. When a cycle has been completed in the game, the results and current state of the information board can be displayed on the blackboard, and a general discussion of the results can be stimulated by the lecturer who can ask such questions as, "This is the present state of the process, what is your assessment of the situation? What would you do next?"

It is perhaps worth emphasizing to the class that the psychological effect of an early EVOP success is very great and makes it much easier to introduce later schemes and to overcome the inertia which usually exists. Some care is needed, therefore, in choosing processes for early EVOP study. Evolutionary Operation has been successful in helping with the solution of some extremely intractable problems; however, it might be unwise to begin EVOP with the most intractable problem available!

Once a few successful EVOP schemes have been run, it is wise to invite the engineers and the process superintendents responsible for these successes to lecture to future classes. This has the advantage of introducing realism and making it clear that EVOP is not some theoretical idea but rather a practical tool that can be put to immediate use. It is particularly useful to have the past users of EVOP discuss their difficulties and how they have overcome them. The first reaction of a busy process superintendent may be that *his* own process is too complicated and difficult for anything as simple as EVOP to be useful or even feasible. He is more easily convinced when he sees how his colleagues have managed to run successful programs in spite of difficulties such as lack of segregation of the product and the presence of large analytical errors.

Plant Operators

About half a day should be enough for instructing the plant operators. The instruction should be carried out by supervisory personnel who have successfully completed the more comprehensive training course for supervisory personnel and who, preferably, have had some actual experience with EVOP. Evolutionary Operation should be presented to the operators as a good idea that needs their cooperation to make it work. It will be explained that, instead of running the plant at fixed conditions, a specified pattern of variants will be used, and their suggestions are invited for deciding on these variants. Visual aids are important, for example, large color diagrams of the EVOP pattern, the plant, the EVOP board with results entered, and so on. Operators need not be told about the details of the calculation procedures, but merely that these procedures provide a way of better understanding the results and of ensuring that chance happenings do not mislead.

In the last resort, the successful carrying out of the EVOP program depends on the plant operator. It is very important that his interest is stimulated and maintained; one of the ways in which this can be done is to have the information board—or at least a replica of that part of it that shows the current average responses—prominently displayed where he can see it, for example, in the process control room. One large company employs a device they have nicknamed the "sputnik" when running EVOP

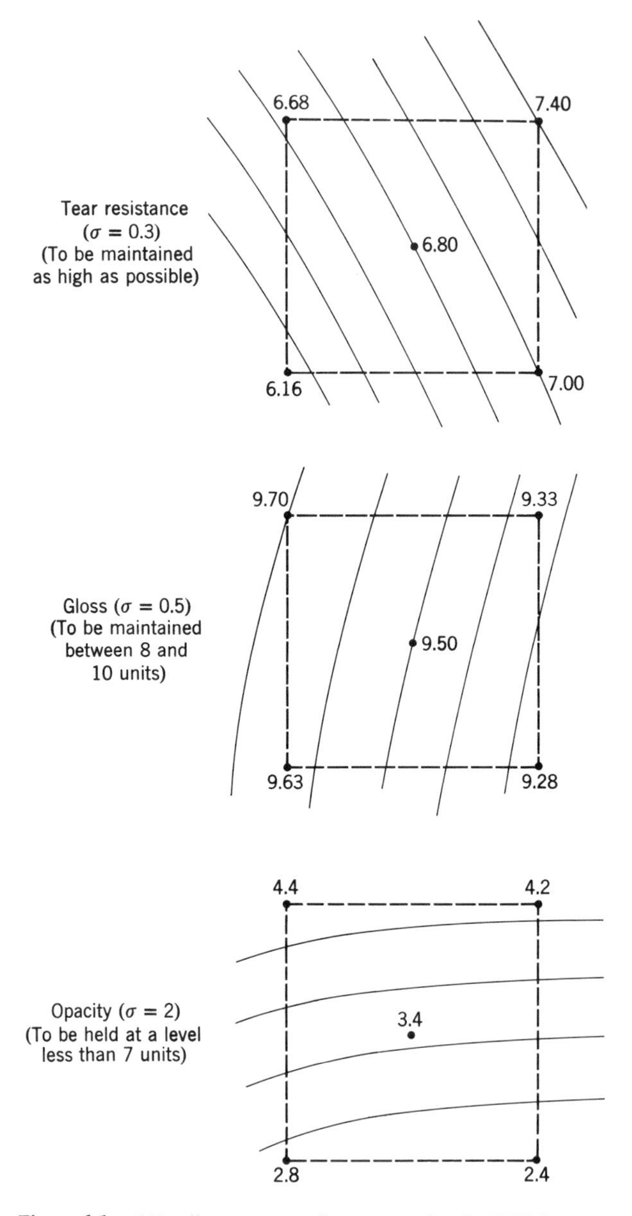

Figure 6.1. *"True" contours and responses for the EVOP game.*

on three variables. This device consists of steel balls connected by rods to form a cube plus center point configuration. A card bearing the current mean response can be attached to each ball which itself represents a certain set of operating conditions. "Sputnik" was kept up to date in the operating room and created considerable and sustained interest. (See photograph.)

Summary

We are concerned, then, with supplying information and training to three types of people having distinct functions in the company. Each presentation has a different objective and should follow the general lines previously described. Considerable effort and care are justified in planning and carrying into effect this educational program, because only through it can the full benefits of EVOP be achieved.

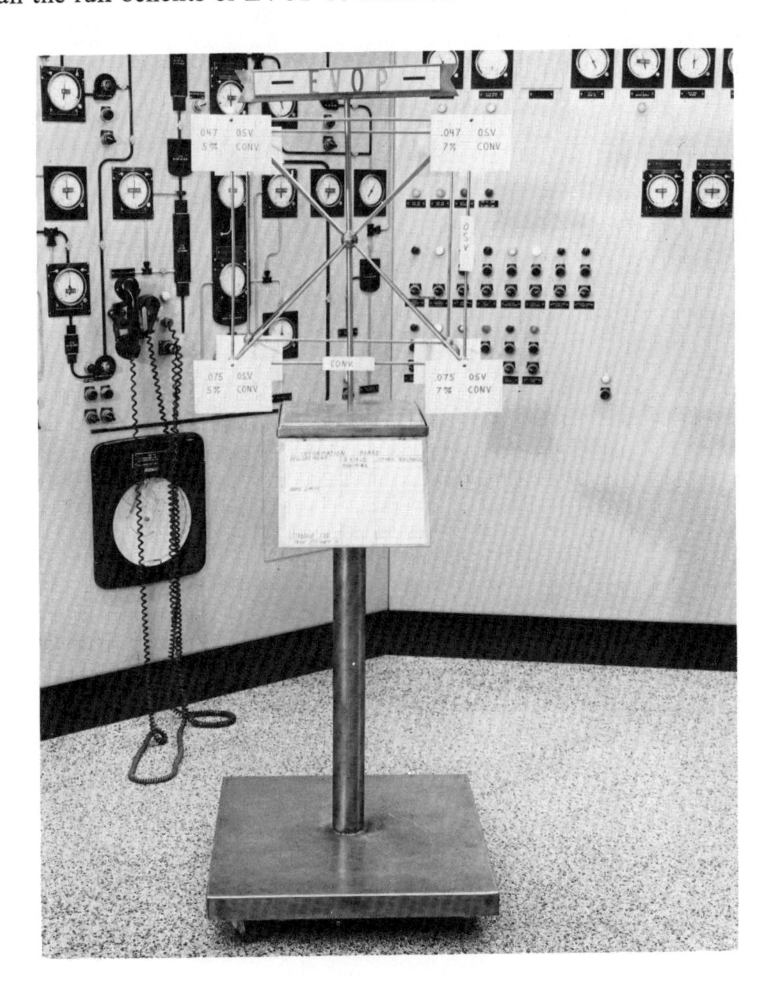

Table 6.1. An actual set of values for the EVOP game

y_1 = tear resistance

y_2 = gloss

y_3 = opacity

4 2 0 1 3

Sample size: 40

Point 0			Point 1			Point 2			Point 3			Point 4		
y_1	y_2	y_3	y_1	y_2	y_3	y_1	y_2	y_3	y_1	y_2	y_3	y_1	y_2	y_3
7.0	10.1	6.2	6.3	10.0	1.9	7.8	9.7	3.4	7.1	9.4	2.1	6.5	9.7	4.0
6.7	8.8	2.2	6.1	10.6	2.3	7.5	8.0	3.0	7.0	8.9	1.8	6.8	9.1	1.0
7.0	10.0	4.1	6.4	9.6	2.8	7.1	9.0	2.9	7.5	8.9	1.7	6.9	9.5	3.4
6.5	7.3	4.0	6.4	10.0	3.7	7.3	7.9	3.4	6.9	8.9	1.8	6.6	10.1	2.8
6.4	9.7	3.5	6.9	10.1	2.7	7.2	9.5	6.7	6.6	8.3	0.0	6.9	9.5	8.2
7.5	10.0	5.2	6.6	10.4	2.9	6.9	9.8	4.4	7.9	9.5	0.2	6.3	9.4	5.7
6.9	8.8	2.4	5.9	9.5	2.8	7.4	10.3	0.8	7.3	8.9	0.0	6.8	10.9	3.9
6.3	9.9	5.4	6.1	9.9	3.1	7.2	8.2	3.4	7.1	9.2	3.7	6.2	10.6	6.5
6.4	9.6	1.9	6.7	10.1	4.2	7.4	8.5	3.6	7.1	8.4	1.6	6.4	9.9	3.1
6.8	9.8	7.6	6.4	9.2	1.2	7.3	9.0	4.3	7.1	9.3	3.6	6.6	9.4	4.3
6.9	8.6	4.6	6.1	9.6	1.8	7.1	9.2	10.2	6.5	9.4	0.0	6.3	9.5	4.6
7.3	8.6	4.8	6.3	9.1	2.3	7.4	8.6	4.9	7.3	9.7	0.3	6.2	9.6	4.0
6.7	9.5	4.3	6.1	10.0	0.6	7.0	10.1	2.6	7.0	9.3	2.3	6.1	9.7	3.5
6.5	9.6	0.8	6.3	9.4	1.4	7.2	9.9	7.1	7.4	10.1	0.0	7.4	10.3	5.3
6.9	9.7	0.7	6.4	9.2	1.8	7.7	9.4	5.1	7.0	9.3	2.9	6.9	10.1	3.5
6.5	8.8	0.0	5.7	9.4	0.6	7.2	9.7	6.9	7.1	8.5	4.0	6.7	10.2	4.6
6.9	9.2	2.3	6.0	9.8	1.0	7.3	9.1	5.7	7.2	8.3	3.8	7.0	10.0	4.0
7.3	10.6	2.9	5.8	9.8	5.4	6.9	9.1	2.0	6.8	9.8	1.6	6.5	9.9	4.0
7.2	9.5	2.1	6.5	9.6	4.1	6.9	9.2	7.1	7.5	8.8	0.2	7.0	9.5	3.5
7.0	9.4	3.4	6.1	9.2	3.5	7.6	10.1	0.4	7.2	8.9	1.4	6.4	9.7	4.4
6.1	9.7	0.2	6.0	9.4	0.8	7.7	8.5	4.1	6.7	9.2	4.4	6.6	10.3	7.5
7.6	10.2	5.6	5.9	10.2	0.6	7.6	9.1	2.2	6.7	9.1	2.8	6.5	8.8	0.0
6.7	9.5	2.6	6.2	10.5	5.0	7.3	9.3	5.7	6.6	9.9	1.2	6.8	9.8	6.4
7.3	10.1	0.0	6.5	10.2	2.3	7.5	8.7	3.0	7.3	9.7	0.9	7.0	9.5	5.6
7.0	8.8	5.2	5.9	9.7	1.6	7.8	8.6	3.6	7.1	9.6	1.4	6.4	9.6	4.9
6.7	9.2	4.2	6.2	9.7	2.7	7.6	9.2	1.9	7.0	9.1	0.6	7.1	10.1	3.2
7.1	9.5	5.1	5.8	9.6	3.0	7.6	8.9	4.4	7.0	9.6	0.8	6.6	9.4	3.1
6.9	9.6	0.0	6.2	9.9	6.4	7.5	9.6	4.0	7.0	8.1	3.3	7.2	9.5	8.1
6.9	10.0	3.0	5.8	9.7	5.0	7.3	9.4	3.4	6.9	8.7	2.4	6.9	9.9	3.9
6.7	8.8	2.6	5.7	9.8	3.0	7.1	9.2	8.4	6.9	9.8	1.3	6.5	9.9	1.4
7.0	10.2	5.0	6.2	9.4	3.6	7.0	8.8	5.2	7.0	9.1	1.3	6.9	9.9	2.5
7.1	8.9	3.1	6.5	9.5	4.4	7.3	9.5	3.9	6.9	9.1	0.0	7.2	10.0	2.0
6.4	10.2	1.8	6.0	10.2	3.1	7.1	9.2	4.5	7.2	9.5	1.4	7.2	9.8	4.1
7.5	9.3	3.6	6.6	9.5	1.2	7.3	8.5	3.9	6.9	9.4	5.1	6.1	9.5	1.9
6.8	10.0	4.5	6.0	9.6	0.2	7.3	10.0	3.3	6.6	9.3	4.1	6.4	9.5	4.0
6.5	10.2	3.1	6.2	9.7	0.0	7.5	10.0	7.8	6.7	9.0	5.6	5.9	10.2	6.3
6.8	8.8	3.3	6.7	10.0	2.0	7.4	8.9	4.0	7.2	10.2	1.9	6.9	9.1	5.7
6.6	8.9	3.7	5.8	8.7	1.9	7.1	9.7	5.5	6.8	8.5	3.4	6.4	9.4	7.4
7.1	9.0	2.6	6.5	9.2	0.8	7.5	10.1	2.7	6.6	8.6	1.4	6.7	9.8	7.1
6.5	9.6	5.2	6.2	10.3	1.8	7.3	8.8	1.8	7.0	10.5	5.9	6.8	10.2	5.0

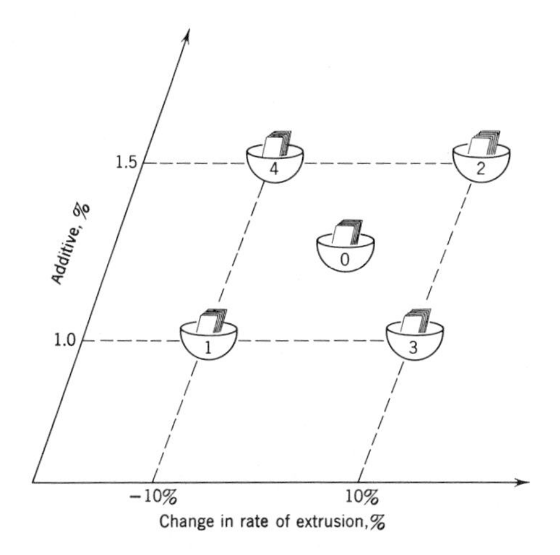

Figure 6.2. Bowls containing counters arranged for the EVOP game.

6.2. SIMULATION OF TWO-VARIABLE EVOP: THE EVOP GAME

The EVOP simulation game is a device of considerable value in the training courses for supervisory personnel. This game is carried out in class immediately after the ideas and principles of EVOP have been explained. The reader can develop his own particular version of the game to suit any special needs, and he may base one on some process familiar to his intended audience, if he so desires. We present here a particular version of the game that closely represents a situation encountered in one application of EVOP. The example given is for a two-variable EVOP program, although the idea could easily be extended to three-variable EVOP.

The simulation game is based on a problem connected with the extrusion of plastic film where two variables, rate of extrusion and the amount of an additive, were being studied. Three responses were recorded—the film's tear resistance, its gloss, and its opacity. The main object of the investigation was to increase tear resistance as much as possible while maintaining specification on gloss and opacity. Specifically, it was desired to maintain gloss between eight and ten units and opacity less than seven units. In the simulation we have supposed that the actual situation was that shown in Figure 6.1. The contours indicate true relationships which might exist between the three responses and the two process variables, and the numbers indicate the true responses which would arise at the five design points. These diagrams are not shown to the student, at least not until the game is over. Their object is only to produce a basis for generating the data.

To represent data as they would actually come from the operating process, five bowls each containing 40 counters were arranged before the class in the frame of reference of the design, as shown in Figure 6.2. Each counter had written on it a value for the tear resistance, the gloss, and the opacity. Using a table of random normal deviates [e.g., RAND Corporation (1955) or Table VII] in the manner described in Appendix 2, the distributions of the counters in the bowls were arranged so that the theoretical mean values and standard deviations for the three responses were those given in Figure 6.1. The actual values used on one such set of counters are given in Table 6.1. Because of sampling variation, the mean values and standard deviations *of the 40 numbers on the counters* are, of course, slightly different from the theoretical ones.

In the actual conduct of the simulation game, a member of the class draws out one counter from each of the five bowls, thus producing the results for the first cycle of the scheme. These are recorded, the five counters are then returned to their appropriate bowls, and the contents of each bowl are reshuffled or mixed before the next set of counters is drawn. The class is divided into three sections, each section working on one response. The results of the first cycle are entered by the class on the worksheets and the appropriate calculations are performed. The complete results from all three responses are set out on the information board (which

Response	Tear Resistance	Gloss	Opacity
Requirement	Raise	8–10	Below 7
Means	6.9 7.1 6.3 6.5 6.7	9.1 9.2 9.9 9.5 9.1	5.7 8.4 5.4 4.4 2.8
2 S.E. limits for means	±1.0	±0.8	±4.4
Phase mean	6.7	9.4	5.3
Effects with 2 S.E. limits: Extrusion	0.2 ± 1.0	−0.2 ± 0.8	0.6 ± 4.4
Additive	0.4 ± 1.0	−0.2 ± 0.8	3.4 ± 4.4
$E \times A$	0.0 ± 1.0	0.2 ± 0.8	2.2 ± 4.4
Change in mean	0.4 ± 0.9	−0.5 ± 0.7	−0.1 ± 3.9
s			
Prior estimate of σ	0.5	0.4	2.2

Figure 6.3. *The information board from an EVOP game—cycle 1.*

Response	Tear Resistance	Gloss	Opacity
Requirement	Raise	8–10	Below 7
Means	7.0 7.0 6.5 6.4 6.6	9.6 9.0 9.7 9.7 9.2	3.8 6.8 4.8 5.4 3.4
2 S.E. limits for means	±0.7	±0.6	±3.1
Phase mean	6.7	9.4	4.8
Effects with 2 S.E. limits: Extrusion	0.2 ± 0.7	−0.5 ± 0.6	0.5 ± 3.1
Additive	0.6 ± 0.7	−0.2 ± 0.6	0.9 ± 3.1
$E \times A$	−0.2 ± 0.7	0.0 ± 0.6	2.4 ± 3.1
Change in mean	0.2 ± 0.6	−0.3 ± 0.5	0.0 ± 2.8
s	0.2	0.4	1.7
Prior estimate of σ	0.5	0.4	2.2

Figure 6.4. The information board from an EVOP game—cycle 2.

Response	Tear Resistance	Gloss	Opacity
Requirement	Raise	8–10	Below 7
Means	7.0 7.1 6.8 6.2 6.8	9.7 9.2 9.6 9.7 8.9	3.9 6.8 4.4 4.6 3.6
2 S. E. limits for means	±0.4	±0.5	±1.5
Phase mean	6.8	9.4	4.7
Effects with 2 S.E. limits: Extrusion	0.4 ± 0.4	−0.6 ± 0.5	1.0 ± 1.5
Additive	0.6 ± 0.4	0.2 ± 0.5	1.3 ± 1.5
$E \times A$	−0.3 ± 0.4	0.2 ± 0.5	2.0 ± 1.5
Change in mean	0.0 ± 0.4	−0.2 ± 0.5	0.2 ± 1.4
s	0.4	0.5	1.3
Prior estimate of σ	0.5	0.4	2.2

Figure 6.5. The information board from an EVOP game—cycle 3.

Response	Tear Resistance	Gloss	Opacity
Requirement	Raise	8–10	Below 7
Means	6.8 7.2 6.9 6.2 6.9	9.6 9.4 9.4 9.6 8.8	3.4 5.8 4.6 4.5 3.1
2 S.E. limits for means	±0.4	±0.5	±1.5
Phase mean	6.8	9.4	4.3
Effects with 2 S.E. limits: Extrusion	0.5 ± 0.4	-0.5 ± 0.5	0.5 ± 1.5
Additive	0.4 ± 0.4	0.3 ± 0.5	0.8 ± 1.5
$E \times A$	-0.1 ± 0.4	0.4 ± 0.5	1.9 ± 1.5
Change in mean	-0.1 ± 0.4	0.0 ± 0.5	-0.4 ± 1.3
s	0.4	0.5	1.5
Prior estimate of σ	0.5	0.4	2.2

Figure 6.6. *The information board from an EVOP game—cycle 4.*

Response	Tear Resistance	Gloss	Opacity
Requirement	Raise	8–10	Below 7
Means	6.7 7.3 6.9 6.3 6.9	9.6 9.4 9.2 9.6 8.7	3.8 5.0 4.6 3.7 3.1
2 S.E. limits for means	±0.4	±0.4	±1.5
Phase mean	6.8	9.3	4.0
Effects with 2 S.E. limits: Extrusion	0.6 ± 0.4	-0.5 ± 0.4	0.3 ± 1.5
Additive	0.4 ± 0.4	0.4 ± 0.4	1.0 ± 1.5
$E \times A$	0.0 ± 0.4	0.3 ± 0.4	0.9 ± 1.5
Change in mean	-0.1 ± 0.3	0.1 ± 0.4	-0.5 ± 1.4
s	0.4	0.4	1.7
Prior estimate of σ	0.5	0.4	2.2

Figure 6.7. *The information board from an EVOP game—cycle 5.*

is usually the classroom blackboard) for all the class to see. There is then a short discussion as to whether or not any useful conclusions can be drawn at this stage.

The actual course of one such EVOP game is shown in Figures 6.3 to 6.7, which give the state of the information board after 1, 2, 3, 4, and

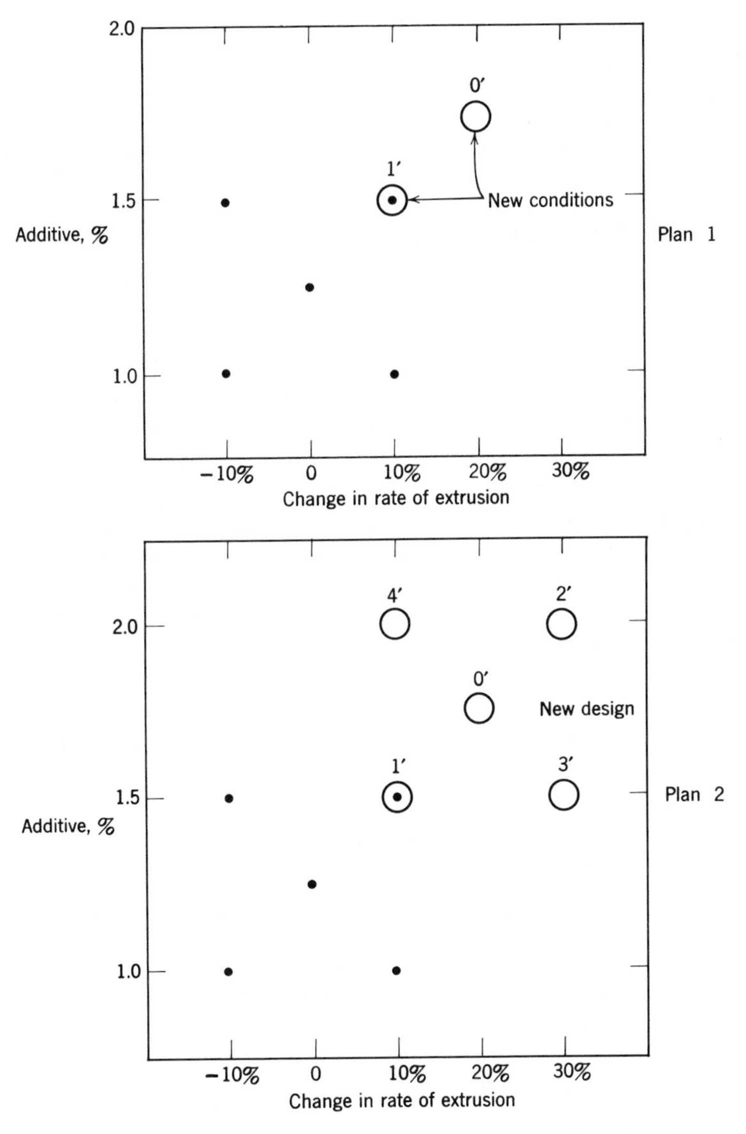

Figure 6.8. *Two alternative proposals for further EVOP exploration.*

Table 6.2. Observations drawn during one specific EVOP game

	Tear Resistance					Gloss					Opacity				
Cycle	0	1	2	3	4	0	1	2	3	4	0	1	2	3	4
1	6.3	6.5	7.1	6.7	6.9	9.9	9.5	9.2	9.1	9.1	5.4	4.4	8.4	2.8	5.7
2	6.7	6.2	7.0	6.6	7.2	9.5	9.9	8.8	9.3	10.0	4.3	6.4	5.2	4.1	2.0
3	7.5	5.8	7.2	7.2	6.9	9.3	9.6	9.7	8.3	9.9	3.6	3.0	6.9	3.8	3.9
4	7.0	6.5	7.5	7.1	6.1	8.8	9.6	10.1	8.4	9.5	5.2	4.1	2.7	1.6	1.9
5	6.9	6.5	7.6	6.8	6.3	8.6	9.2	9.2	8.5	9.4	4.6	0.8	1.9	3.4	5.7

5 cycles have been performed. All numbers were rounded to one place of decimals. The specific observations drawn from the bowls are shown in Table 6.2. After each cycle the class discussed whether they felt that some change was justified or whether a further cycle of observations should be taken. For this particular set of drawings most of the class were not in favor of modifying the process until the fourth cycle, at which point a majority concluded, correctly, that the process conditions corresponding to 1.5% additive with a 10% increase in rate of extrusion were the best conditions actually run, and that further exploration in the direction of increasing rate of extrusion with the additive at higher levels was indicated. However, since careful consideration had to be given to meeting the specifications on gloss and opacity, a majority said they would run at least one further cycle to be somewhat more certain of the effects of the process variables on the secondary responses, i.e., gloss and opacity. After viewing the results of the fifth cycle, the whole class was in favor of exploration in a north-east direction, increasing both rate of extrusion and percentage of additive. There was some discussion as to the best way to do this. One group favored a proposal to pursue this direction initially by alternating the following conditions (as indicated in Figure 6.8, plan 1):

Change in Rate of Extrusion, %	Additive	New Point Number
+10	1.5	1′
+20	1.75	0′

A smaller group favored a proposal to begin immediately a new pattern of variants hinged on the north-east corner of the previous pattern. This second proposal is also shown in Figure 6.8 as plan 2, with the new points indicated by 0′, 1′, 2′, 3′, and 4′.

After some deliberation the class finally decided in favor of plan 1 on the grounds that it was more flexible. If the conditions 0′ give the expected improvement in tear resistance with no gloss or opacity problem, then we might quickly extend the probe to 2′ and *possibly further*. The location of the new region to be explored by a new design might thus end up further out in the north-east direction than that shown in plan 2 or less far out if difficulties were encountered.

The objective of the EVOP game is not only to give supervisory personnel practice with the calculations as well as familiarity with the worksheets but also to come reasonably close to the actual conditions of EVOP, and to illustrate the considerations that would have to be borne in mind in evaluating the information board at any given stage. This particular example was constructed by taking an actual EVOP example and, after the principal effects had become clear, constructing a population of suitable size having approximately the means and variances that actually occurred. The reader should have no difficulty in constructing other examples of this type by using a table of random normal deviates to generate the populations. Details are given in Appendix 2.

6.3. AIDS TO SUCCESSFUL EVOP

The EVOP Committee

The existence of the EVOP committee ensures that results are adequately reviewed. It also ensures that a list of ideas for further study is always available and is continually up-dated. The reviewing process itself frequently generates new ideas. One phase of EVOP uncovers promising leads which are followed up in later phases.

This regular, periodic, technical review of the process in the light of data so far generated ensures that the responsible process superintendent has brought to his attention, at regular intervals, a clear summary of the operation of his process at the current standard levels and at modified levels. He does not have to make his examination alone, but has available the help and advice of other experts who can often suggest ideas and interpretations which he might not see himself. As we have already mentioned in Section 1.9, these other experts might be:

1. A research chemist or chemical engineer who may possess some specialist knowledge of the particular type of process under study or, more generally, may simply have the proper technical background, even if he is not familiar with the specific process.
2. A representative from the quality control department who can provide answers to questions involving quality or specifications.
3. A statistician who may review certain statistical aspects of the problem in excess of the normal EVOP statistical requirements.

We now discuss the contributions of these members of the EVOP committee in more detail.

The Research Man's Contribution. In the review of an EVOP program, data on the chemical and physical properties of the principal product and of by-products are considered. The statistical analysis on the information board reveals the changes that occur in all these properties when specified changes in the operating variables are made and may suggest, to a person familiar with the basic chemistry of the process, an explanation of what is happening. These conjectures will frequently give rise to ideas to be explored in further EVOP phases. These ideas may involve new ways to manipulate variables already considered or they may involve quite new variables. It frequently happens that a single clue leads to a whole new sequence of fruitful investigation along lines not previously thought of at all. Sometimes clues to what is going on in the process cannot be conveniently followed up on the full scale. The research man, however, provides a link to other facilities. In some instances, possibilities will have sufficient potential to lead him to recommend investigation in the laboratory or on the pilot plant.

The Quality Control Man's Contribution. Suppose that it turns out that the percentage yield can be improved only at conditions which change the texture of the product to such an extent that this characteristic will be outside specification limits. We must then ask the question: "Is the present specification in texture one which we must meet to satisfy the customer or is it one laid down long ago on some arbitrary basis or for reasons that are now irrelevant?" We need to have someone at the meeting who can answer such questions or can get an answer. He will be responsible for seeing that any changes in specifications that result are acceptable to all concerned and that the new specifications are officially instituted.

The Statistician's Contribution. A useful addition to the EVOP committee is a statistician or at least a person with deeper statistical knowledge than is usually obtained from brief training in EVOP. The statistician can help to interpret more difficult statistical questions and in some instances where, for example, several variables produce complicated interactions, he may occasionally suggest the use of a more complex statistical program employing the ideas of response surface methodology [see Davies (1956); Cochran and Cox (1957)]. He should be listened to, but at times it may be necessary to resist his proposals. Care should be taken to see that the central idea of EVOP as a simple procedure capable of being carried out as a routine by process workers is not lost. Elaborate investigations, where they are needed, should be conducted by special teams of investigators, and such investigations should not be treated as part of the EVOP program.

Recognition of Success in EVOP

One basic principle of good company organization is to make the good of the company and the good of the individual synonymous, so far as this is possible. The conduct of successful EVOP should be given fair recognition and reward, otherwise excuses may be found not to continue. A process superintendent may be deterred if he is afraid that the *only* reaction of management to successful EVOP will be automatically to raise production requirements to a new level. Some form of tangible recognition is needed to encourage efforts that succeed in raising productivity.

A Constant Flow of New Ideas

Evolutionary Operation is an alternative way of running a process; it is not a temporary switch to experimental manufacture. The question asked should be, "What are we running on EVOP this month?" rather than, "Are we running EVOP this month?" The lifeblood of successful EVOP is a constant stream of new suggestions for process improvement, suggestions from all company levels. Two sources of such ideas are:

1. The EVOP committee who will be continually stimulated by the actual results coming from the process itself.
2. The process operators who, if encouraged, frequently make excellent suggestions.

The primary reason for failures in EVOP programs is a lack of imagination. For example, we sometimes hear, "EVOP is all very well, but this process has only two really important variables, temperature and pressure, and these have been thoroughly investigated already." A statement of this kind is usually unjustified. It may mean that, from some particular view of the process, these are the variables which we would *expect* to have major effects and this is why they have been the only ones examined. Such a view is usually found to be too narrow. The introduction of variations of a type not previously contemplated will often produce effects which can be exploited in one way or another. The typical process operating manual will usually contain many pages specifying temperatures, pressures, holdup times, concentrations, agitation speeds, rates of temperature and pressure buildup, and so on. If the question is asked: "Are we certain that these are the best settings?" the honest answer would usually be "No, we are not certain. These figures are based on small scale work and on some hastily conducted startup runs. Very few of these settings have been carefully checked out on the full scale." Questions of this nature should continually be asked and new ideas for EVOP should constantly be sought.

Education: A Secondary Aspect of EVOP

A secondary but important aspect of EVOP is its value as a means of installing basic statistical ideas in industry. Among the ideas are the following:

1. Observations *vary*.
2. Variation can be *measured*.
3. We can *distinguish* between apparent effects of change which *can be fully accounted for by background noise* and others that cannot.
4. Process variables can be deliberately *rocked* or *pulsed* and made to yield information on their effects.
5. By changing or pulsing variables simultaneously in an appropriate pattern or design, we can get information about *several variables*, instead of only *one*, with equal efficiency.
6. These designs are simple to understand and bring to attention the importance of *interaction*—a concept previously unfamiliar to many engineers.
7. By the simple process of blocking, it is possible to get rid of a great deal of systematic variation which would otherwise impede progress.
8. There is great advantage in thinking of problems geometrically, particularly when we are dealing with multiple responses.

(*a*) Thus, we can think of a *space* with coordinate axes on which are marked levels of the variables under study and we can visualize the contours of a response surface lying within this space. If these contours were known, they would tell us the level of the response for any selected levels of the variables under study.

(*b*) We can conceive of a run as a *point* in the space.

(*c*) When we make a run, we can imagine putting a *probe* in the space at some point and determining the response there.

(*d*) By carefully selecting our pattern of probing spots, we can get an idea of the shape of the response surface.

These modes of thought alone are of great value and can, of course, be extended with profit to other industrial problems besides those that arise immediately in an EVOP program.

Now it is true that, to instill these and other concepts, we could arrange that all process engineers take a formal course in statistics over a period which might cover a number of months. Some companies regularly do this. However, if this is *all* that is done, the results are sometimes disappointing. The members of the class have all kinds of other things to worry about in their day-to-day work, and topics are quickly forgotten and frequently never come to life. In particular, the student often never learns how the various tools fit together.

Now, an EVOP course need occupy only four half-days in which the student is taught just the basic fundamentals. He learns something about variation, the 2 S. E. limits, design and the 2^2, 2^3 factorials. However all this forms a package and the student can *apply* that package almost at once. The package provides a rather rigid framework designed to keep him out of trouble. It shows one way, at least, in which statistical ideas can fit together. His hand is being held. The design and the way he will use and analyze it are all preprogrammed so far as this is possible.

A very important element in learning is commitment, but commitment is dangerous. EVOP supplies the engineer with a simple package that he can apply himself without much danger of running into trouble. It has built into it all the elements for success which statistical design and analysis can devise. The effect of this, when done right, is that the man will gain a great deal of confidence. Whatever he finds out from his first runs of EVOP, he has been, as it were, in the firing line. He has made deliberate changes in the process in accordance with a statistical pattern and has analyzed the results. The difference between this man and one who has just taken a statistical course is the difference between a veteran soldier who has been in the war and one who has merely drilled on the square. It is true that the theoretician knows much more about statistics than he does, but he has something the theoretician does not always have—experience and motivation.

Quite a few of the people who run an EVOP program become enthusiastic. They were taught the minimum amount of statistics and design, but they have applied it and it has come to life before their eyes. Such people, are usually anxious to hear more and they will have the confidence and motivation to apply what they hear.

Specifically the ideas learned in EVOP about measurement of variation can form a basis for further study of quality control, confidence intervals, significance tests, time series, and forecasting. The ideas concerning the use of 2^2 and 2^3 designs can be used to introduce more elaborate factorials with blocking, fractional factorials, and other designs. The geometrical ideas lead naturally to consideration of response surfaces and regression analysis. The ideas about balancing responses to achieve specifications produce a basis for the study of linear and nonlinear programming. Finally, and not the least important, is the human factor. Many companies have special-process development groups, experimental groups, trouble shooters, operations research workers. To be useful these groups must be staffed with people who are imaginative and enterprising. How are people of this kind found? One way is to seek out those who have run successful EVOP programs.

CHAPTER 7

EVOP, Optimization, and Variations of *EVOP*

7.1. INTRODUCTION

The Iterative Nature of Experimentation

A number of modified versions of EVOP have been proposed from time to time. To discuss these properly, we first explain what objectives we feel an EVOP scheme should set out to achieve. Basically, such a scheme is a method of learning more about a process by inducing the process itself to supply the relevant data. To understand what is being attempted, we must consider the process of learning embodied in any scientific investigation. Investigation is necessarily an iterative procedure, as shown in Figure 7.1. Initially the investigator will have some conjecture in mind. Thus, for example, in an EVOP program it might be conjectured (1) that changing certain variables in certain directions would have a beneficial influence on the process. A series of runs is then planned (2); data become available (3); and these data are then analyzed (4). It should be understood that the word *analysis* does not refer solely to the formal statistical analysis of results but to every aspect of their fuller consideration which can lead to a modified conjecture and to the commencement of a further iterative cycle aimed at illuminating the situation further. After a few cycles of an EVOP phase, it may appear that changes in particular variables are accompanied by changes in certain responses. The investigator may, as a result, decide on one of a number of alternative courses, for example, to rerun the design with different levels of the variables, perhaps following an apparent trend in response levels, or to rerun with one or more new variables, and so on. Whatever the decision made, the point is that we are led to our *next* conjecture through speculation illuminated by the results from all previous

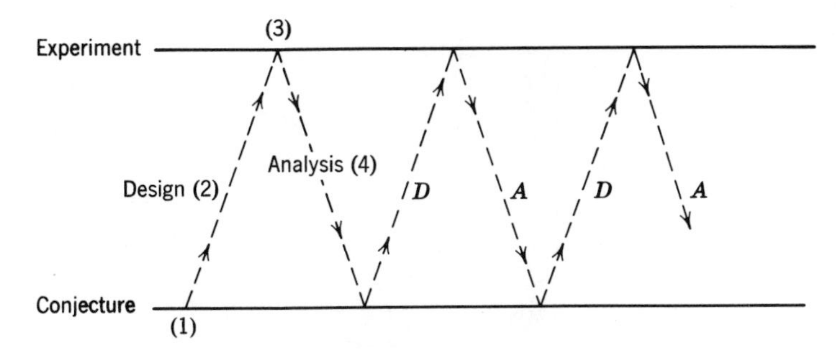

Figure 7.1. The iterative nature of experimentation.

stages. Thus our learning procedure is *iterative*. We see that a *feedback* process is involved. In the case of EVOP, the results are being fed back to the process superintendent and the EVOP committee who are then taking appropriate action.

Empirical and Scientific Feedback

It is important to distinguish between two kinds of feedback:

1. Empirical (or *routine, automatic, idiot*) feedback.
2. Scientific (or *technical*) feedback.

Empirical feedback is of the type where a particular response pattern leads, more or less automatically, to a particular action. Thus, for example, we might run the process at two different temperatures and proceed with whichever temperature provided the higher yield. Such feedback requires no scientific or technical explanation of what is going on, and undoubtedly does play an important role in EVOP.

A more important role in EVOP is played by scientific feedback. In this type of feedback, the results interact with technical knowledge to produce actions which could not be taken on a purely automatic basis. The principal purpose of the EVOP committee is to ensure that scientific feedback does take place. As a result of careful study of the chemical and physical effects which occur, clues are often provided about what is going on in the process. Such clues suggest new variables and new types of changes which were not initially contemplated. We are thus led along a path largely inaccessible to empirical feedback alone.

An Example of Scientific Feedback

Figure 7.2 shows the average values for three responses related to polyethylene plastic film obtained after four cycles of an EVOP program to-

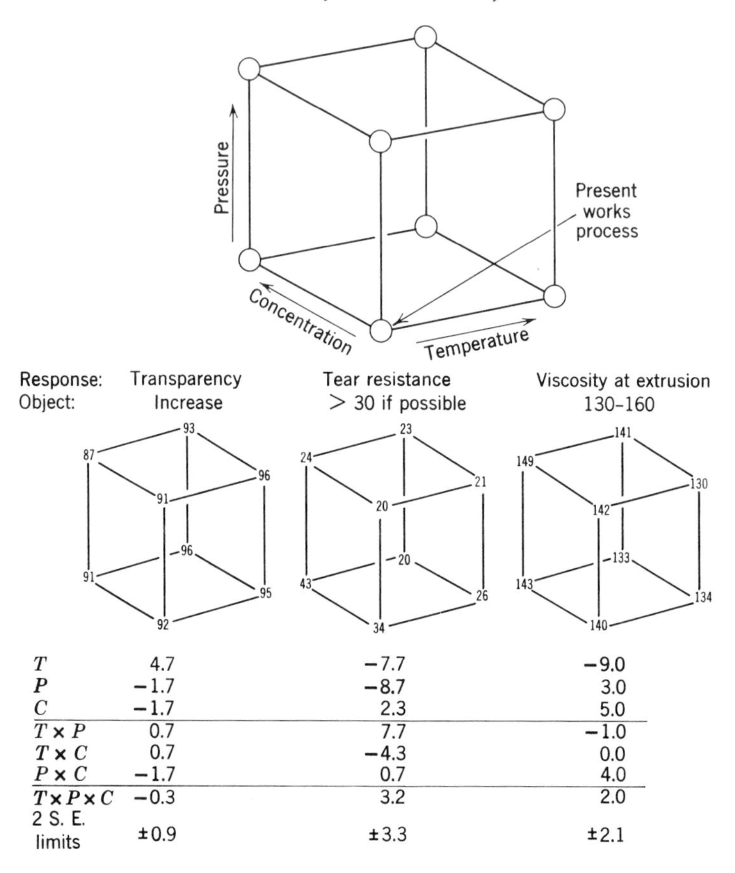

	Transparency	Tear resistance	Viscosity at extrusion
T	4.7	−7.7	−9.0
P	−1.7	−8.7	3.0
C	−1.7	2.3	5.0
$T \times P$	0.7	7.7	−1.0
$T \times C$	0.7	−4.3	0.0
$P \times C$	−1.7	0.7	4.0
$T \times P \times C$	−0.3	3.2	2.0
2 S. E. limits	±0.9	±3.3	±2.1

Figure 7.2. *A three-variable EVOP phase after four cycles.*

gether with all the possible factorial effects and their 2 S. E. limits. In this particular phase, the effect of increasing temperature, concentration, and pressure above their standard values was being studied. The principal objective was to obtain high transparency of the extruded film. It is clear from the transparency diagram that high transparency is favored by a high temperature so that, if high transparency were the only consideration, a change to higher temperature would be adopted. Such a change would be an example of empirical feedback.

In the practical situation, it was essential that not only should the transparency be high but also that the *tear resistance of the film should be high* and, if possible, greater than 30. Unfortunately, the changes that increase transparency evidently reduce tear resistance to an unsatisfactory extent.

It would, of course, be possible to compromise. For example, some weighted combination of the two criteria might be developed to supply an over-all "figure of merit." Weighting might be based, for example, on monetary value of increased transparency and of increased tear resistance. Conditions might then be chosen which maximize this figure of merit. Action taken because of such an approach would again be the result of empirical feedback, although of a little more sophisticated kind. In the situation actually pictured, however, because of the basic conflict that exists, it is obvious that very little improvement is to be expected as the result of any compromise.

One of the other quantities recorded was viscosity. A specification between 130 to 160 on this property was introduced to ensure the proper operation of the extruding machine and, since all the recorded values fell within these limits, it did not at first attract much attention. It was then noticed that the factorial analysis for viscosity and the factorial analysis for transparency were very closely related. The estimated effects for transparency were very nearly proportional to those for viscosity, but with opposite signs. This was not only true for main effects but for moderate-sized interactions as well. At this point a plot was made of transparency against viscosity and, as was to be expected from the factorial analysis, a very strong relationship was found (Figure 7.3). It was now pointed out by the technical people that a theoretical explanation could be advanced supporting a simple causative connection between viscosity and transparency. Such a theory would imply that, under the particular conditions of opera-

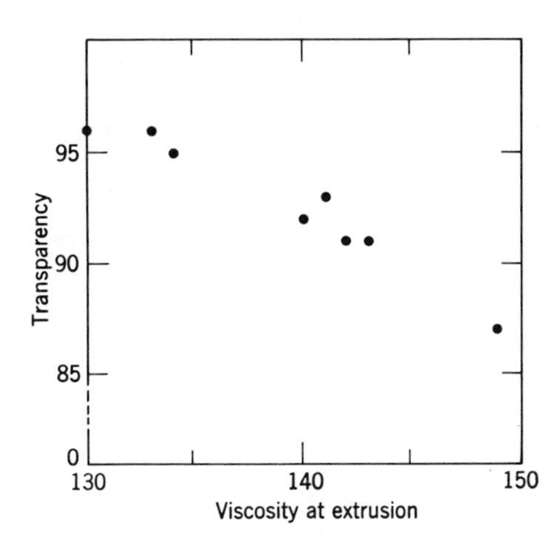

Figure 7.3. A plot of transparency against viscosity.

tion employed, transparency was directly dependent upon the viscosity at extrusion, and that changes in transparency which resulted when temperature, pressure, and concentration were changed occurred *only* because these three factors changed the viscosity. If this theory were correct, then presumably other factors which reduced viscosity would also increase transparency. The problem was to find some such other factor which produced no undesirable corollaries and, in particular, did not decrease tear resistance. A number of possibilities were tried, and fairly soon a chemical additive was found that performed the desired function.

The action taken in this instance was as a result of scientific feedback. The resulting discovery, it will be noted, was not of a trivial character, since it overcame a fundamental barrier to empirical development of the process.

The Different Consequences of Empirical and Scientific Feedback

The reason why the distinction between empirical and scientific feedback is important is because the circumstances which ensure efficient scientific feedback are not necessarily those which ensure efficient empirical feedback, and vice versa. For efficient scientific feedback we need a basis from which we can safely reason about the results we seem to see. We need enough replication to obtain estimates of adequate reliability. By adequate reliability we mean that the standard errors of the effects are sufficiently small so that we can at least be reasonably sure of the sign of the true effects, and for some purposes considerably greater precision would be needed.

Thus for scientific feedback *repetition* of the EVOP cycle and consideration of the size of the effect in relation to the 2 S. E. limits in the manner we have proposed makes sense. However, if we were concerned only with empirical feedback, such a procedure might *not* be very good. In fact, we show that most efficient *empirical* feedback may occur when there is *no* replication at all and no consideration of 2 S. E. limits.

Models for Empirical Feedback and Their Consequences

In considering optimal feedback strategy, we must employ a model which takes account of the following basic fact. To spend two months rather than one month in testing a particular modification is a decision which denies the opportunity to test, in the second month, some second modification which could possibly have a much more beneficial effect than the first. It is possible that more can be gained by testing more variants than by increasing the certainty that the usefulness of a given modification is correctly decided.

A model which takes this fact into account is discussed by Box (1966*a*). Although this model is simple, it is probably sufficiently realistic to be

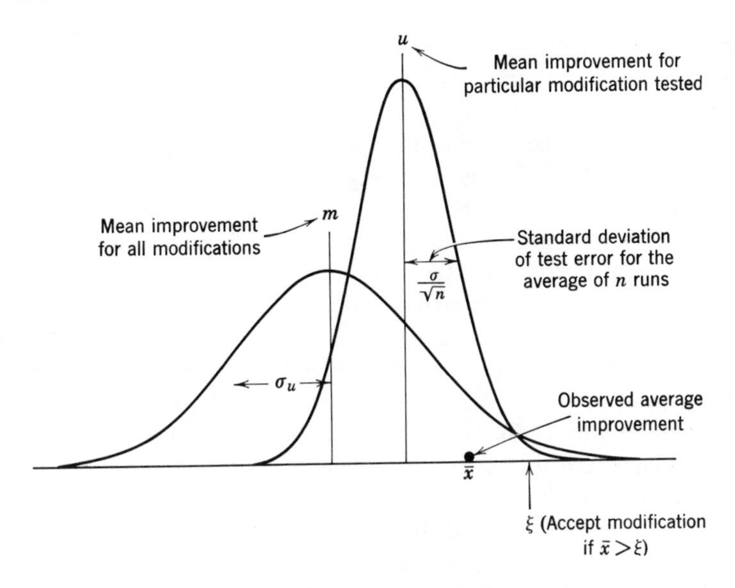

Figure 7.4. An empirical feedback model.

useful. It is supposed (see Figure 7.4) that numerous alternative process modifications are available that might be tried in an EVOP scheme. The act of choosing a modification is imagined to be paralleled by drawing at random from an urn that contains tickets bearing names of all the possible modifications. It is further supposed that the true effect u of any such modification is measured in terms of *the increased profit per unit of time* obtained from its use. The true effects u for the population of possible modifications form a distribution which we shall suppose to be normal with some particular mean m and standard deviation σ_u. A modification with unknown true effect u is drawn randomly, and it is now supposed that n runs are used to test this modification. Let $\bar{x}$ be the average improvement in response which is an estimate of the effect of the modification selected. The over-all test error is supposed to be normal with a variance σ^2 so that $\bar{x}$ is distributed about u with the standard deviation $\sigma/\sqrt{n}$. The modification is accepted if $\bar{x} > \xi$ and discarded otherwise, where ξ is some preselected value. We may ask what value should be chosen (a) for ξ, the critical value for the apparent average improvement, and (b) for n, the number of observations on which $\bar{x}$ is based.

Another quantity that we need in the derivation is the extent of the remaining life of the process. We can imagine, for example, that a particular chemical process on which EVOP was being run might have a future life of perhaps 500 weeks. In such a case we could test 500 modifications for 1 week each or 5 modifications for 100 weeks each. Clearly, if we test each modification very thoroughly, we shall be more certain that the decision to accept or reject the modification is correct. On the other hand, if we spend a great length of time in testing each modification, then we forego the possibility of testing other modifications which might be better. The balance should be made by choosing the values ξ and n so as to maximize the total expected profit over the remaining life of the process.

On the basis of the model we have discussed, it turns out that the optimal choice of ξ and n heavily depends on the mean m of the effect distribution. Now the value m determines whether or not the modifications to be tried produce, on average, an improvement ($m > 0$), a harmful effect ($m < 0$), or no difference ($m = 0$). Now, when we raise the question of what m might be in a practical industrial situation, we find a good deal of disagreement. Some people believe that we should take $m > 0$. They argue that because the modifications available to be tried have been carefully screened and selected, most of them have a better than average chance of being of value. Others argue that, because the process has already been studied in some detail, most of the possible modifications will be harmful, leading to the assertion that $m < 0$.

A compromise might be to set m equal to zero, thus supposing that the chance of choosing a favorable modification is equal to the chance of choosing an unfavorable modification. In this situation, it can be shown that, under the *empirical* feedback rule previously outlined, the maximum total profit is obtained by setting $n = 1$ and $\xi = 0$. Thus, *however large the testing error*, the modification will be tested *once and once only*. If the test result x is greater than zero (the single test shows an apparent improvement), the modification will be adopted; otherwise, it will be rejected.

If m is set greater than zero, it turns out that the critical value ξ should now have a negative value. However, the optimal value of n is still unity. *Only if m were negative could replicate testing be justified.* For details, see Appendix 3, Section A3.1.

The reason for these somewhat surprising results is that, with $m \geq 0$, although many of the modifications which are going to be accepted when there is no replication will make the process worse, nevertheless a number of these modifications will improve the process. Moreover the net beneficial effect will be, on average, greater than the beneficial effect of the smaller number of "good" modifications which could have been selected as a result of more careful and extended testing.

An alternative empirical feedback model which has a different basis but again leads to the conclusion that only a single run should be made at a given set of conditions can be formulated as follows. Suppose a continuous variable such as temperature is measured on a scale such that zero corresponds to the present temperature, and values . . . , -3, -2, -1, 0, 1, 2, 3, . . . represent convenient steps up and down the temperature scale. Suppose a response such as yield is locally linearly dependent on temperature, and we wish to increase this response as much as possible. To do this we might use the following procedure:

We run n experiments at the value 0 and n experiments at the value 1, and if the average response at 1 is higher than the average response at 0, we move up one step on the temperature axis and repeat the procedure or, otherwise, move down one step and repeat it. It is natural to ask how large n should be so that, after completing a certain total number of runs, the highest possible response is achieved on the average. Once more it may be shown that maximum progress is made, on the average, by setting $n = 1$, no matter what the standard deviation of the observations. A justification of this result is set out in Appendix 3, Section A3.2.

The Empirical-Scientific Feedback Dilemma

For efficient *empirical* feedback simplified arguments thus lead to the conclusion that, in plausible circumstances, a modification should be included or rejected as a result of a single test, whatever the size of the experimental error standard deviation σ. On the other hand, unless experimental error is very small, to obtain efficient *scientific* feedback *we need replication.* Only thus can sufficient accuracy be gained for speculation to be profitable about the possible *meaning* of the effects.

In considering this dilemma the following should be remembered:

1. Unless σ is very small so that a single replication will be enough to obtain a reliable estimate of an effect, we have no way of knowing which of the modifications introduced will have improved the process and which made it worse. Thus, although we may have optimal empirical feedback, no basis will exist for scientific feedback.

2. Unless σ is very small, the single-replication strategy, although the *best* for empirical feedback, may nevertheless not be very *good.* That is to say, if σ is not small, progress may be very slow, even though it is the most rapid that can be achieved with empirical feedback alone. This is because a substantial proportion of the modifications introduced actually make things worse, and this proportion increases as σ increases. The use of scientific feedback, on the other hand, can lead to major improvements inaccessible to empirical feedback.

3. Single replication is seldom a practical strategy if, as almost invariably happens, we have auxiliary responses which must be maintained within

limits. This is because, unless the standard deviations of these auxiliary responses are small, we have no way of knowing if the inclusion of the modification will yield a product which violates the constraints in the auxiliary responses or not.

4. Experience with many EVOP programs shows that a large number of major improvements have occurred as a result of scientific feedback which has opened up new and fruitful avenues of investigation.

Our conclusion, therefore, is that, primarily, we should learn more about the process. To do this there must be adequate replication, and designs which make ready interpretation possible should be used. On the other hand, the fact that each additional cycle we run in a given phase is denying us the possibility during that cycle of studying other variables must be continually borne in mind, and an attitude of controlled impatience must be cultivated. The investigator should be continually pushing to get on with the program, to try new variables, and should not get bogged down.

Fortunately, the divergence between the requirements of empirical and scientific feedback is not as wide as one might fear. We show in Appendix 4 that, when the ranges of the variables are well chosen, three or four cycles of an EVOP scheme based on a 2^3 design can provide a good chance of revealing at least the sign of important effects.

7.2. OPTIMIZATION METHODS AND EVOP

Evolutionary Operation has sometimes been referred to as an *optimization* technique and is compared with other optimization techniques. Such comparisons must be considered with some caution.

Evolutionary Operation is an experimental technique for seeking the *preferable*. By contrast, problems of mathematical optimization* are usually concerned with determining the maximum of some function $z = f(x_1, \ldots, x_k)$ within some region $R(x_1, \ldots, x_k)$ of the space of the x's, where usually it is assumed that:

1. The variables $x_1, \ldots, x_k$ are known.
2. The region $R(x_1, \ldots, x_k)$ is known.
3. The nature of the function $z = f(x_1, \ldots, x_k)$ is known.
4. The function $z = f(x_1, \ldots, x_k)$ can be computed without error for any chosen set of x's.

"Best" strategies of mathematical optimization attempt to determine the maximum of the function with the smallest amount of effort. It should

* One important class of such problems consists of what are called *linear programming problems*. Here it is supposed that z is a linear function of $x_1, \ldots, x_k$ and that the region $R(x_1, \ldots, x_k)$ can be defined by linear inequalities. [See Hadley (1962); see also Section 7.3, "Linear Programming."]

be noted that, even when we can make the foregoing assumptions, to seek an over-all best strategy is probably pointless. The nature of the function being investigated and the nature of the optimal strategy for that surface are two sides of the same coin. For example, if the effects of the factors involved were approximately additive, so that the function was of the form $f(x_1, \ldots, x_k) = a_0 + a_1 f_1(x_1) + a_2 f_2(x_2) + \cdots + a_k f_k(x_k)$, where the a's are constants, then a fairly simple strategy *which assumed additivity* would be effective. However, most functions found in practice are not of this form and often a more complex strategy would be needed. In particular, we often may not know enough initially about the properties of the function to know how to choose the strategy. If two persons A and B play a game in which A first specifies a strategy and B then specifies a surface on which to use that strategy, it is clear that, *whatever* strategy A suggests, B could almost always find a function which would defeat it. For this reason, we cannot speak about one strategy of optimization being best in any sort of general sense, though we might speak of a strategy being "best" for some particular class of functions.

Even if we consider more sophisticated strategies which provide for learning about the function and for modification of tactics as we proceed, we must still build into our learning procedure our prior expectations of the kinds of functions we might expect to encounter.

The situation which is met when we attempt to improve an industrial process using EVOP differs from the mathematical optimization setup in a number of respects:

1. We do *not* know the variables $x_1, \ldots, x_k$ which should be included in the function f.
2. We do *not* know with exactitude the region $R(x_1, \ldots, x_k)$ in which we should seek to maximize z.
3. We do *not* know the functional form f.
4. The observations are usually subject to *moderate or large errors.*

It should be clear that procedures applicable for the mathematical optimization problem may be totally unsuitable for the EVOP situation. In fact, it is dubious whether optimization is a useful concept at all for the latter. We can talk of preferable conditions but rarely, if ever, of optimal conditions.

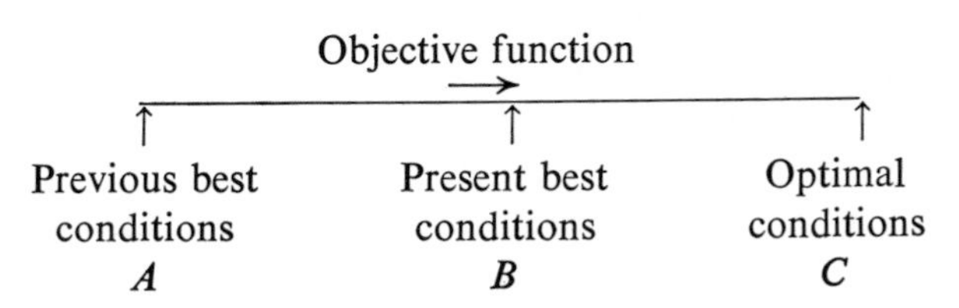

In attempting to maximize an objective function such as the profitability of the process, we could say that the present conditions B were *preferable* to the previous conditions A, and the difference $B - A$ was a measure of the progress which *had* been achieved. By contrast, the optimal point C would often be an entirely nebulous concept.

We can know, at any given stage of an investigation, that we have found conditions which are *preferable* to those previously used. We can rarely know that we have attained *best* conditions. Some of the reasons are as follows:

1. Even if a process existed where improvement of a *single* response such as yield was the only criterion, we would seldom know how far short of this theoretical point was the practical maximum.
2. In practice, the objective function will almost always need to take into account a number of responses such as yield, cost, purity, color, physical form, and so on. It would be even less possible to be sure that the optimum objective had been attained when improvement was so many faceted. Moreover, the importance of the various characteristics will change from time to time depending on the state of the market, the nature of competitive products, and so on. It is extremely unlikely that we could ever be sure that the ultimate for a multidimensional and shifting objective function of this kind had ever been achieved.
3. It might be argued that, even though we seldom know *a priori* the highest value which an objective function z can attain, we can nevertheless be fairly sure that at least we have reached a *local* maximum by study of the behavior of the function close to this supposed stationary value. Although true, this is of less help than first appears. We may have reached a local maximum in the space of some set of studied variables $x_1, x_2, \ldots$. Unfortunately, we never know *all* the variables that ought to be included in the set $x_1, x_2, \ldots, x_k$. For example, suppose, in a batch reaction the technique employed in the past allowed the reagents to react at a fixed temperature for a fixed time at a fixed pressure. We have mentioned three variables—reaction temperature x_1, reaction time x_2, and reaction pressure x_3. Experimentation may lead us to discover best levels of these three variables, perhaps a temperature of 160° with a reaction time of $2\frac{1}{2}$ hours and a pressure of 60 psi. We have then, let us suppose, a maximum in the space of these variables x_1, x_2, and x_3.

Suppose now that a laboratory investigation suggests that impurity formation will be lower if pressure is not maintained at the same level throughout the reaction but is slowly raised from a low value to the higher level of 60 psi. The pressure versus time profile under consideration is now that shown in Figure 7.5*b* rather than that in Figure 7.5*a*, and two new variables are being contemplated which were not there before. They are the initial pressure and the time to achieve full pressure, which we have denoted in the diagram by x_4 and x_5.

The dimensionality of the problem has changed and what may possibly have been an absolute maximum in the space x_1, x_2, x_3 is unlikely to be an absolute maximum in the space of x_1, x_2, x_3, x_4, and x_5. Obviously the process of increasing the dimensionality of the problem is unlikely to stop here. How do we know that the general shape of profile given in Figure 7.5*b* is really the best? Perhaps, towards the end of the reaction, pressure ought to be reduced again. If so, by how much and for how long? What about temperature? Perhaps a nonuniform temperature-time profile would be better.

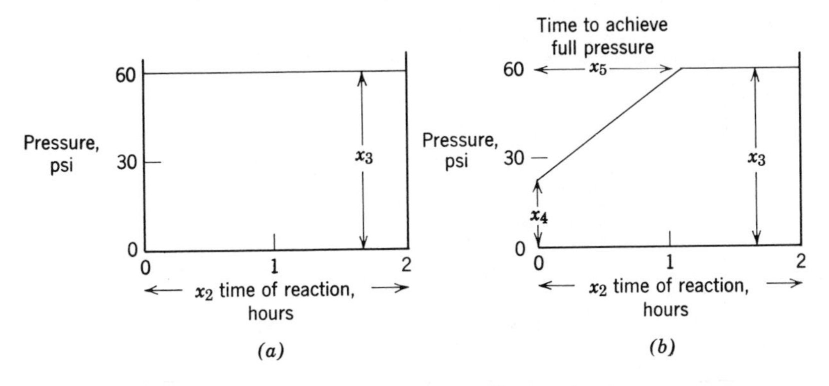

Figure 7.5. Pressure versus time profile for a batch process.

Typically, as an EVOP program develops, the dimensionality of the experimental space develops in ways that could not have been predicted at the beginning of the investigation, and there is little point in discussing EVOP problems as if the space of the variables were fixed.

4. Not only are the variables $x_1, \ldots, x_k$ unknown but also the region $R(x_1, \ldots, x_k)$ in which it is possible to work is often unknown and, to some extent, flexible. If, in exploring a direction of improvement, we meet what has, for some reason, been considered a boundary in the space of the x's, we naturally ask if that boundary can be moved, and often it can.

In summary, then, we see that in the typical EVOP situation we cannot say at any point that no new possibilities for improvement will become available. The optimum is like the pot of gold at the end of a rainbow.

7.3. SOME OPTIMIZATION TECHNIQUES RELATED TO EVOP

We have seen, then, that EVOP has many aspects. It is a method of process improvement that:

1. Can be readily conducted under actual process conditions.
2. Can operate in circumstances where the important variables are not known and, therefore, has a screening or selecting aspect.
3. Allows adaptive scaling. Appropriate scaling for the variables will not usually be known. If the scales are badly chosen in one phase, they can be adjusted in the next phase to give a wider or narrower span between levels.
4. Can operate in the presence of experimental error.
5. Can provide an efficient basis for scientific feedback.
6. Does not assume knowledge of the functional form of the response nor, in fact, any explicit knowledge of the response function except that it is "smooth."

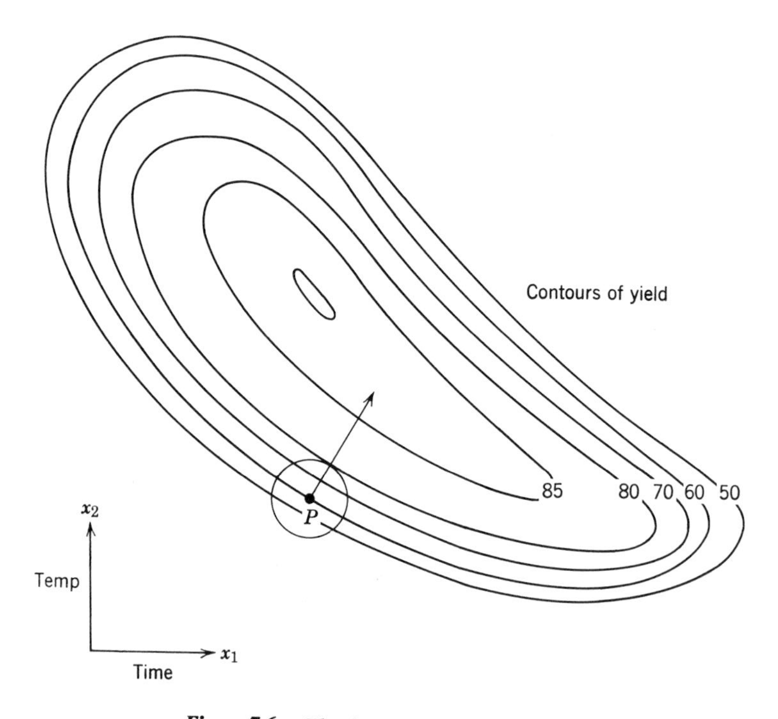

Figure 7.6. *The direction of steepest ascent.*

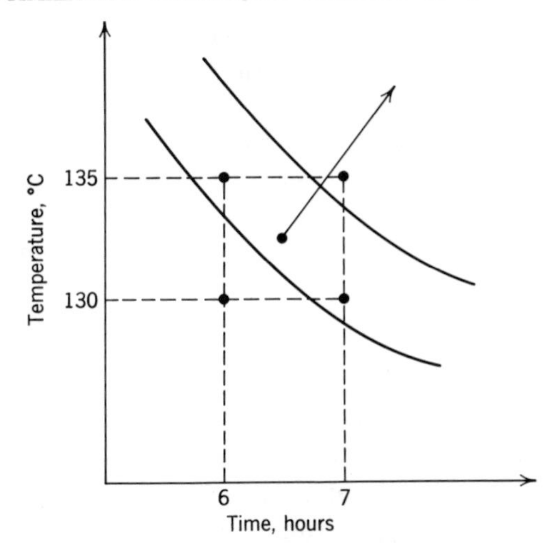

Figure 7.7. A steepest ascent design with the scales of the variables shown.

We now briefly discuss some related optimization techniques.

Steepest Ascent

Figure 7.6 shows contours of a maximal region for a response y depending on variables x_1 and x_2. In an actual experimental problem these contours would usually not be known, though in some problems the equations of the contours would be known. It is our object to move from some initial point P in the (x_1, x_2) space toward the point of maximum response in the center of the contour system. Imagine a small circle drawn about P, as in the figure. Then consider the directional line obtained by joining the center of the circle to the point on the circumference of the circle where it just touches one of the response contours. As the diameter of the circle is decreased, this directional line is said to point in the *direction of steepest ascent* on the surface at the point P. This direction achieves the greatest rate of increase per unit of distance traveled in the space. It can easily be shown that this direction is followed by making, in the variables x_1 and x_2, changes which are proportional to the derivatives $\partial y/\partial x_1$ and $\partial y/\partial y_2$, evaluated at P respectively. The method has often been used for obtaining the maximum or minimum of known functions. Its use as an experimental procedure where the function is unknown but observations, subject to error, can be taken was introduced by Box and Wilson (1951); see also Davies (1956). In this experimental procedure a small experimental design is performed around the point P (see Figure 7.7) and the derivatives

$\partial y/\partial x_1$ and $\partial y/\partial x_2$ are estimated numerically, from the resulting observations, as the effects of the variables calculated in the usual way.

The method of steepest ascent is not invariant to scale changes and, in its experimental use, we obtain an estimate of the direction of steepest ascent measured in the units used to scale the design. Thus in Figure 7.7 we see that the experimenter has made a change from six to seven hours in the variable time and a change from 130° to 135° in temperature. The direction shown is one of steepest ascent in a scaling where an interval of one hour and of 5°C are represented by equal lengths in the figure. In some other scaling the path would be one of ascent, but not necessarily of steepest ascent. The lack of invariance to such changes is not a serious disadvantage, particularly since the investigator can improve his choice of units as he goes along, just as he learns what are appropriate variables to examine as he goes along. In EVOP we do not recommend the formal calculation of the steepest ascent path, and a more informal procedure seems appropriate. The investigator naturally tends to change variables in the direction in which improvement is expected to occur, and, in doing so, follows an ascent path. Since in EVOP programs there are usually many considerations to be taken into account in deciding the next step, and since in any case only two or three variables are involved, the formal calculation of the direction of steepest ascent is not particularly helpful.

Response Surface Study

Steepest ascent will be most effective as a preliminary procedure to get in the right "ball park" when the region first explored is remote from the maximum. The response surface in such a region will be well approximated by a tilted plane, and a two-level factorial design performed in such a region will yield large main effects which measure the slopes of the plane in the directions of the variables. When through previous application of steepest ascent, or otherwise, a position close to a maximum has been obtained, linear effects will become small, and further application of steepest ascent becomes unprofitable. At this stage a closer estimate of the maximum and some idea of the nature of the surface near the maximum can often be obtained by augmenting the basic factorial design (filled dots) by additional experimental runs (unfilled circles), as illustrated in Figure 7.8, to form what is called a *composite design.*

This design can be arranged so that the original factorial with center points and the additional added points form two "blocks," so that shifts in the general level of response which might occur between the performance of the two parts of the design may be eliminated. The complete design allows an approximating quadratic surface to be fitted to the observations. (In some cases the fit could be improved by first making suitable trans-

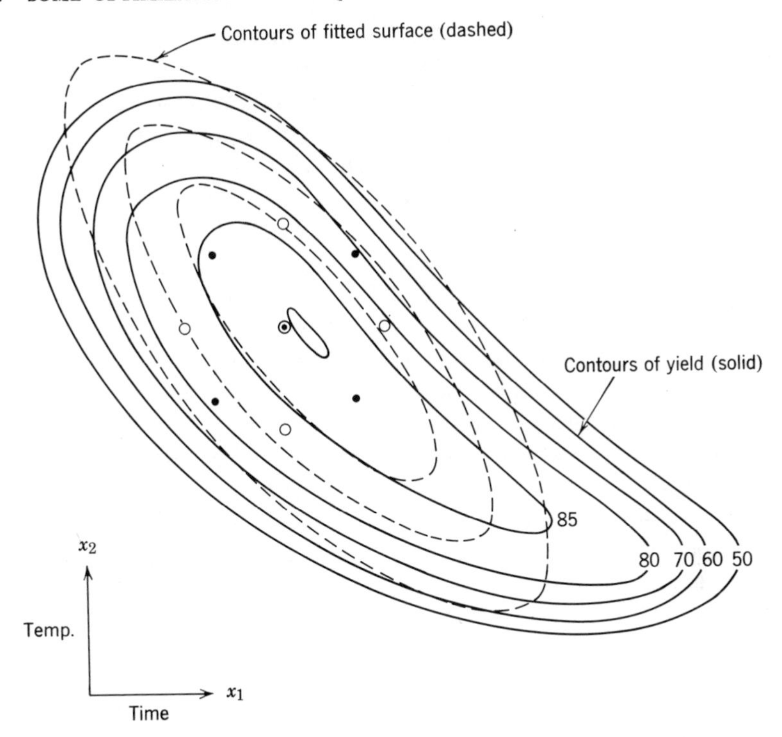

Figure 7.8. Fitting an equation of second degree.

formations in y and the x's.) The contours of such a fitted surface are shown by dotted lines in Figure 7.8. If checks show that the fit is reasonably good, this analysis can (*a*) indicate where final confirmatory runs ought to be made, and (*b*) show the nature of the surface, locally.

In the case illustrated, although the true surface is too complex to be closely fitted by the quadratic surface, the latter reflects the more important features of the true surface. It correctly indicates, for example, that the maximum contains an oblique northwest-southeast ridge. This fact could be made use of if, for some reason, it were essential to use temperatures lower than the optimal temperature. We would know that such a discrepancy from the optimal temperature could be compensated, to some extent, by a longer time of reaction.

The foregoing account is necessarily brief. For a more detailed discussion of response-surface methodology the reader is referred to the original papers by Box and Wilson (1951) and Box (1954*b*), or the accounts in Davies (1956) or Cochran and Cox (1957). This technique would normally be too complicated to be used under the circumstances in which EVOP is employed. However, occasionally, under the guidance of the statistician who will normally attend meetings of the EVOP committee, a composite design of the type shown in Figure 7.8 may be run and response-surface methods may be used to clarify a difficult situation.

Linear Programming

As usually practiced, linear programming is an optimization technique of the specific kind discussed in Section 7.2. There are, however, certain aspects of this procedure that relate to EVOP. For those unfamiliar with standard linear programming, we illustrate with a much simplified example. Suppose we have two additives, 1 and 2, in a gasoline, the amount in tenths of 1% of additive 1, and of additive 2 being denoted by x_1 and x_2, respectively. Suppose that we wish to produce a gasoline with an octane number of not less than 98, and a mile-per-gallon rating of at least 22 in a standard car tested under standard conditions. Various mixtures of the additives 1 and 2 could produce desirable octane numbers and miles-per-gallon ratings. Suppose that the mixtures giving desirable octane numbers are represented by values of x_1 and x_2 such that

$$3x_1 + 5x_2 > 98, \tag{7.3.1}$$

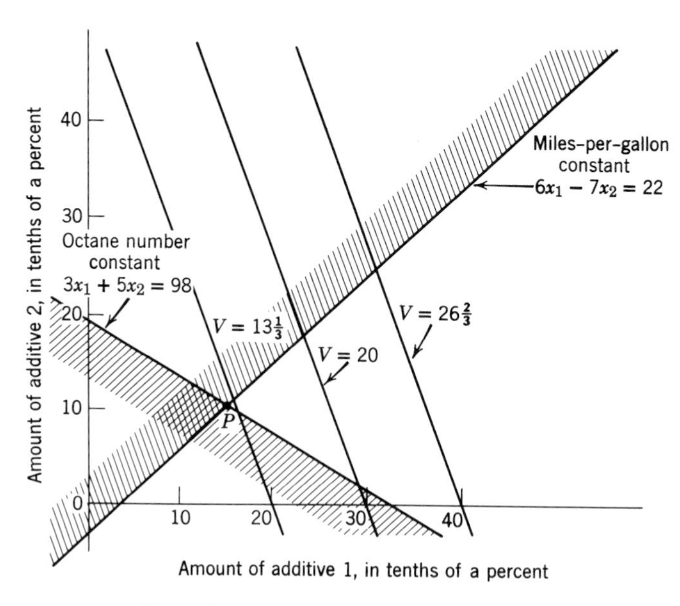

Figure 7.9. *A simple linear programming problem.*

whereas the mixtures which provide desirable miles-per-gallon ratings are given by

$$6x_1 - 7x_2 > 22. \tag{7.3.2}$$

Suppose that the cost V in cents per gallon is represented by

$$\tfrac{2}{3}x_1 + \tfrac{1}{4}x_2 = V, \tag{7.3.3}$$

and we desire to minimize V—the objective function—subject to the constraints (7.3.1) and (7.3.2). We can reproduce this situation diagrammatically by Figure 7.9. The boundaries of the constraints (7.3.1) and (7.3.2) are represented by the two lines

$$3x_1 + 5x_2 = 98,$$

$$6x_1 - 7x_2 = 22.$$

Both constraints are satisfied by any point (x_1, x_2) falling in the region which is to the unshaded sides of both lines. Furthermore, the contours of cost V in cents per gallon are indicated by the lines $V = 13\tfrac{1}{3}$, $V = 20$, $V = 26\tfrac{2}{3}$, and so on. It is obvious from the figure that a gasoline satisfying the requirements and costing the minimum possible is constructed by using the values (x_1, x_2) which correspond to the point P.

This problem of finding the maximum or minimum value of a linear objective function subject to linear constraints such as (7.3.1) and (7.3.2) is referred to as the *problem of linear programming*. The particular example we have given is, of course, trivial. Real linear programming problems will contain many constraints, and these and the objective function will involve a large number of variables. It is this multiplicity of variables that makes the problem a complicated one and from which arises the need for special solution algorithms [e.g., see Charnes, Cooper and Henderson (1953), Vajda (1956), and Hadley (1962)]. It should be noted that, in problems of this type:

1. The objective function and all the constraints are supposed to be linear* in the variables $x_1, x_2, \ldots$.
2. All the functions involved are assumed to be exactly known.

A parallel problem can arise in EVOP. Suppose, for example, that in a particular phase of an EVOP scheme we are varying temperature x_1 and concentration x_2. Suppose our objective is to improve a primary response, i.e., *percentage yield*, but constraints in two secondary responses, *impurity*

* Nonlinear programming problems have also been considered. See, for example, Hadley (1964).

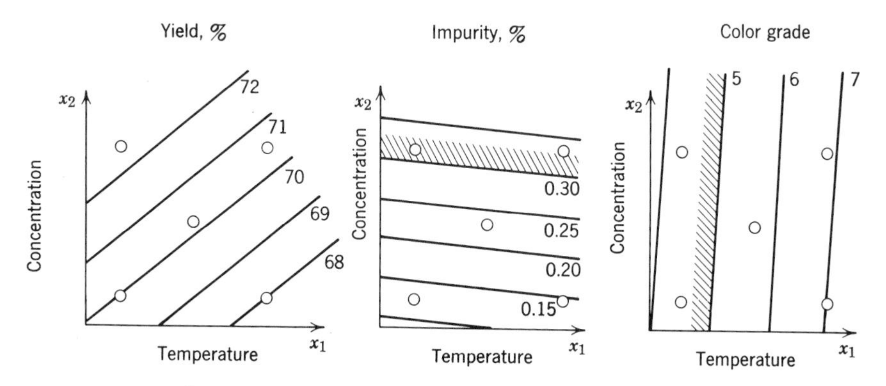

Figure 7.10. *Fitted contours of yield, impurity, and color.*

and *color*, must be satisfied. Suppose, specifically, that we require an impurity level of less than 0.3 and a color grading greater than 5. After a number of cycles of an EVOP scheme, we may be able to draw rough contour lines for the three responses like those shown in Figure 7.10. These allow us to estimate the position of the constraining boundaries in impurity and in color. Figure 7.11 shows the two restrictions on a single diagram, with yield contours representing the objective function also drawn in. In this particular example it will be seen that we can satisfy the constraints and obtain maximum yield at P. We are, in fact, involved in a problem of the linear programming type.

In an experimental EVOP situation the problem is in one sense simpler

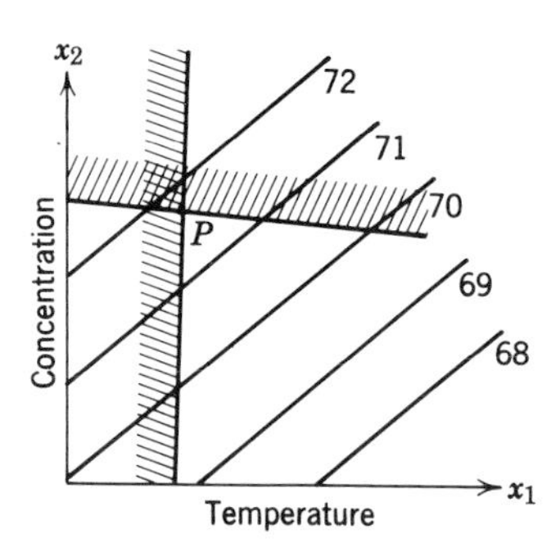

Figure 7.11. *The constraints and the yield contours.*

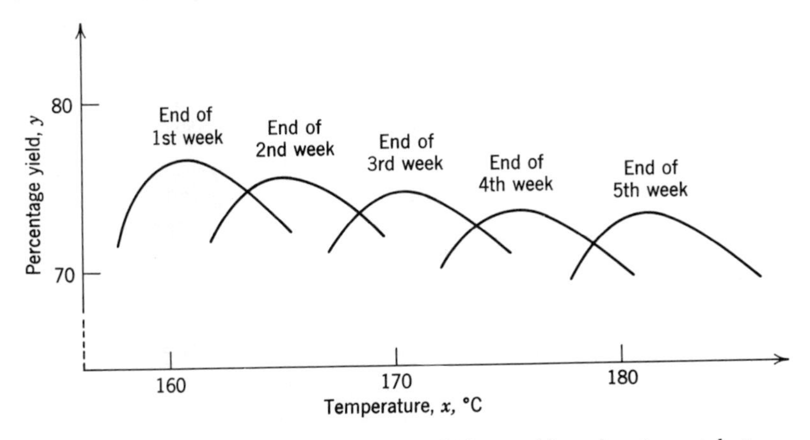

Figure 7.12. A typical response curve drift caused by a decaying catalyst.

and in another more complicated than in the standard linear programming situation. It is simpler because only two or three variables (x's) are involved. It is more complicated because the expressions for the primary response (the objective function in the linear programming setup) and for the secondary responses (the constraints in the linear programming setup) are *not* exactly known mathematical functions but are approximations estimated, with error, from observations.

In the EVOP situation a kind of informal linear programming is often done to give us a rough idea of what kind of changes can be expected to lead to improvement while satisfying constraints. Usually a rough geometrical appraisal is sufficient in EVOP situations since only two or three x variables are involved, and no application of a formal linear programming algorithm is normally necessary. We shall, in any case, regard such an appraisal merely as a guide to provide indications as to where further confirmatory or exploratory runs ought to be made.

Automatic Optimization

Certain types of perturbation techniques for automatic optimization have points of contact with EVOP, and this relationship is also worth considering. Suppose that, in a certain process, the objective function y is the percent yield of a particular product and we wish to maintain this at as high a level as possible. Suppose that yield is affected by the activity of the catalyst, which is renewed every five weeks. It may be known that, because of catalyst change (referred to as *catalyst decay*), one or more of the operating variables may need to be adjusted throughout the six-week catalyst life. It may also happen that no batch of catalyst behaves quite the same as any other, so that the changes needed cannot be forecast in advance.

A typical situation is illustrated in Figure 7.12, where changes in the catalyst are supposedly compensated by the single variable, temperature. We see that the steady decay in the catalyst causes the response vs. temperature curve to "drift" in the direction of increasing temperature. Thus during the course of the catalyst life, higher and higher temperatures are needed to obtain the best yield which itself decreases as the weeks go by. We suppose, as would often be the case, that the exact course and rate of the drift cannot be predicted in advance and will change from occasion to occasion. What we wish to do is to follow the drift, continuously adjusting the temperature to its best possible level.

An approach to this problem is to install equipment which will automatically estimate and make the necessary adjustments as an integral feature of the process [Draper and Li (1951), Box and Chanmugam (1962), Kotnour et al. (1966)]. One method for doing this is to impose a perturbing

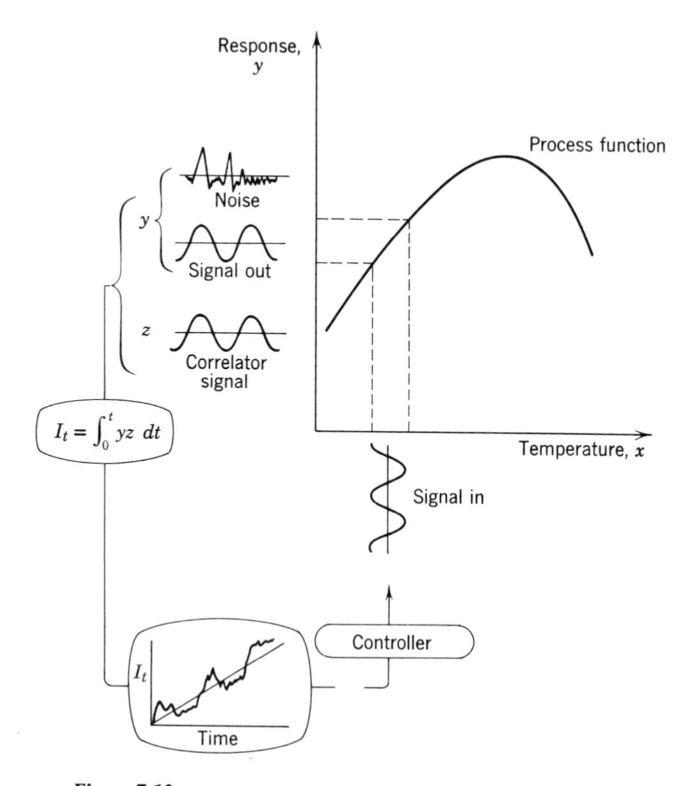

Figure 7.13. Automatic optimization in a continuous process.

waveform onto the variable "temperature" and use the observed variation in response to apply a correction to the *average* temperature level x. The method is illustrated in Figure 7.13. In this illustration the perturbing signal is a sine wave in temperature. The observed response y will then consist of an induced waveform, which is roughly sinusoidal, partially buried in noise.

Suppose, now, that the yield y is continuously measured electrically. Its sinusoidal component may be extracted in the following way. A correlator signal z is generated having the same frequency as the input sine wave. It is also arranged that, when $\partial y/\partial x$ is positive, the correlator signal will be in phase with the signal in the output y. A multiplier and integrator mechanism is now employed which generates the integral

$$I_t = \int_0^t yz\, dt.$$

Now the adjustment that should be made to the average level of the temperature x depends on the sign and magnitude of the derivative $\partial y/\partial x$.

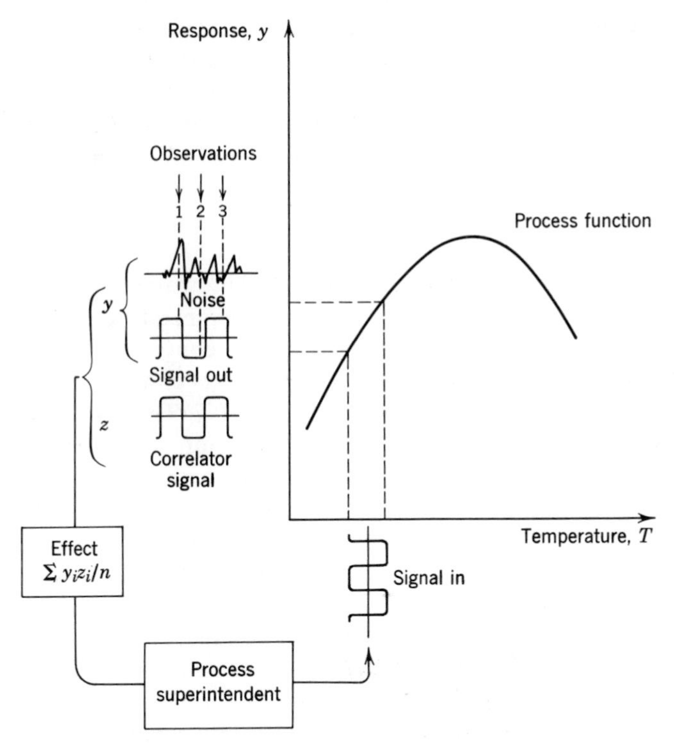

Figure 7.14. *Evolutionary Operation on a continuous process showing empirical feedback.*

But the average rate of change of the integral I_t is proportional to $\partial y/\partial x$, so that the information contained in I_t may be fed back to the controller, which then adjusts the average level of temperature upward if $\partial y/\partial x$ is positive and downward if $\partial y/\partial x$ is negative. The diagram shows the situation with the feedback loop broken after the integrator.

It should be understood that the foregoing discussion is somewhat simplified and, in particular, ignores the dynamics of the process. It is however adequate for the present purpose, and the interested reader may consult the cited references for a more detailed account.

From one point of view EVOP is related to this perturbation technique, as is illustrated in Figure 7.14. If, in a continuous chemical process, we were to carry out an EVOP scheme for the single variable temperature, using two levels run in alternation, we should be introducing the square wave perturbation indicated in the figure. If dynamics were ignored, the output response would contain the square waveform partially buried in noise. In the EVOP scheme we would normally not record the yield continuously but would observe it at discrete intervals. The *main effect* T of temperature on yield would be found by adding together all observations obtained at the upper level of temperature, subtracting all observations obtained at the lower level of temperature, and dividing by the number n of completed square waves fed into the system. Now suppose z_i is a quantity which has the value $+1$ when temperature is at the high level and -1 when it is at the low level. Then the calculation of T is equivalent to

$$T = \sum_{i=1}^{2n} \frac{y_i z_i}{n},$$

in which the numerator directly parallels I_t in automatic optimization. The process superintendent observes the effect of temperature and uses it to adjust the mean level of temperature. He is the "controller" in EVOP.

We see that EVOP does contain certain elements of a perturbation optimization technique. However, this is not all that EVOP is or even what it is principally. A scheme of automatic optimization of the kind we have described uses special and elaborate hardware to control a *known disturbance* (catalyst decay, in this case) through a variable (temperature) whose compensating effect is known. Like other control procedures such methods are, of course, of great value and may be features of processes on which EVOP is employed as an over-all method of investigation.

When a disturbance such as catalyst decay produces a *slow* rate of movement of the response surface, an EVOP-like manual procedure has occasionally been used to stay close to optimum conditions. In one such application the levels of flow rate and temperature in a catalytic reaction were controlled over the (six month) catalyst life by continual repetition of a 2^2

factorial design. The "effects" indicated, at any given time, whether higher or lower levels of the flow rate and temperature should be used at that point in time.

This occasional use of EVOP as a "poor man's adaptive optimization" should not obscure its principal role in screening and examining a whole range of possibly important process variables. Automatic optimization is a perturbation *control* technique, whereas EVOP is principally a perturbation *screening* technique.

7.4. SOME OPTIMIZATION TECHNIQUES UNRELATED TO EVOP

There are a number of other formal optimization techniques which have been suggested principally for functions assumed to be error-free, perhaps with specified functional forms. These techniques can also be employed, less successfully, on functions which *are* subject to error. Among these techniques are those of *random search*, *partan*, and *Fibonacci search*, for example. For a discussion of these, the reader is referred to Wilde (1964). Since the methods are not particularly related to EVOP, we shall not discuss them further.

7.5. SOME SUGGESTED MODIFICATIONS OF EVOP

Various modifications of EVOP have been suggested from time to time.

ROVOP (Rotating Square Evolutionary Operation)

In some of the EVOP schemes with which he was associated in the Monsanto Company, Dr. Edwin C. Harrington found that EVOP com-

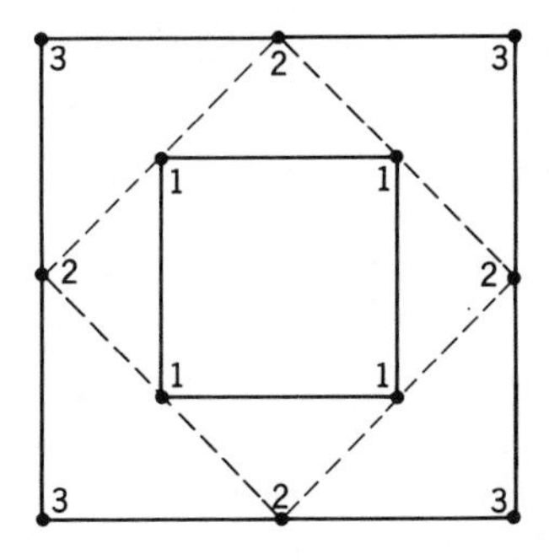

Figure 7.15. Three successive ROVOP patterns.

mittees seemed to be rather too cautious in their choice of the limits between which variables were changed. Because of this, an excessively large number of EVOP cycles might be needed to reveal effects which could have been discovered more efficiently by making somewhat larger changes. A remedy suggested by Dr. Harrington is illustrated in Figure 7.15. Square 1 shows an initial EVOP pattern to examine two variables. If this fails to reveal effects after a few cycles, a new pattern is selected. This second pattern, square 2, consists of a square rotated at 45° from the original square and is of such a size that, in the scales used, the first square just fits into the second. If again no effects are revealed after a few cycles, a third pattern, square 3, is used, obtained in a similar fashion from square 2, as previously described, and so on. When significant effects are seen, the design is moved cautiously in the indicated best direction and the procedure is repeated. For a more detailed account, see Lowe (1964).

Although there are situations where successive modifications of the kind suggested by Dr. Harrington provide possibilities worthy of consideration, especially when an overcautious plant attitude prevails, we feel that, generally, the process superintendent and the EVOP committee should be left free to choose a new pattern in any way they think fit, rather than to follow a preordained strategy. This freedom will, of course, include the possibility of "widening up" the design.

REVOP (Random Evolutionary Operation)

A second EVOP modification, suggested by F. E. Satterthwaite (1959) and discussed by Lowe (1964), involves the use of random points instead of a planned factorial design to provide random EVOP. Part of the intention behind this idea is to allow for the possibility of including a very large number of variables in an EVOP program at one time.

Our own experience suggests that there is very little to recommend this procedure. If EVOP is to be used *routinely* for process improvement and is to be run by plant operators without special supervision, then it is difficult, if not impossible, to make changes in more than two or three variables in the same EVOP phase. Additionally, the random designs proposed by Satterthwaite hinder understanding. The visual appreciation of the effects of the variables on the response is restricted, and the simple comparisons so valuable in factorial experimentation are not available.

Simplex EVOP

An interesting method of optimization has been suggested by Spendley et al. (1962); these authors refer to the method as *simplex Evolutionary Operation*. The basic idea can be understood by considering the case for which there are just two variables x_1 and x_2, and it is desired to improve

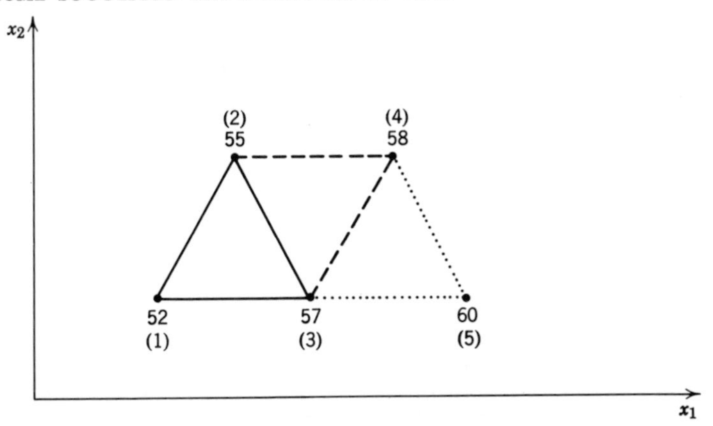

Figure 7.16. Simplex EVOP in two dimensions.

the response percentage yield. The points labeled 1, 2, and 3 in Figure 7.16 show an initial design arranged in the form of an equilateral triangle. The yield results obtained from these three runs were 52, 55, and 57 respectively. Since 52 is the lowest result, a point labeled 4 is added opposite to the vertex 1, and forming a new equilateral triangle with the vertices 2 and 3, and a run is made at these new conditions, giving a yield of 58. The lowest of the runs 2, 3, and 4 is now number 2, namely, 55. The next run is therefore performed at point 5, again opposite the vertex giving the lowest yield value. Runs 3, 4, and 5 now form a third equilateral triangle. Since the yield at point 5 is 60, the next run (6) would be made opposite vertex 3 and positioned so that runs 4, 5, and 6 form an equilateral triangle, and so on. Progress would not always be continuous, of course. Sometimes we would return to a previous vertex, perhaps even cross and recross a baseline between two vertices several times.

In any number of dimensions k there is always a regular figure called a *regular simplex* which contains $k + 1$ points evenly arranged about some center. In two dimensions the regular simplex is an equilateral triangle, as was used previously. In three dimensions (Figure 7.17) the regular simplex is an equalsided tetrahedron. Note that, if we imagine the four vertices of the tetrahedron joined to its center point (C in the figure), then the lines so obtained will make equal angles one with another. Sets of points which form the vertices of regular simplices (plural of simplex) can be written down for any number of dimensions. It is a property of the k-dimensional regular simplex with $k + 1$ points that we can consider any vertex, "reflect" it with respect to the space of the remaining k vertices, and obtain from the reflected point and the remaining k vertices a new regular simplex in k dimensions. (We previously illustrated this in two dimensions.) Thus by dropping the lowest observation of the regular simplex points at each

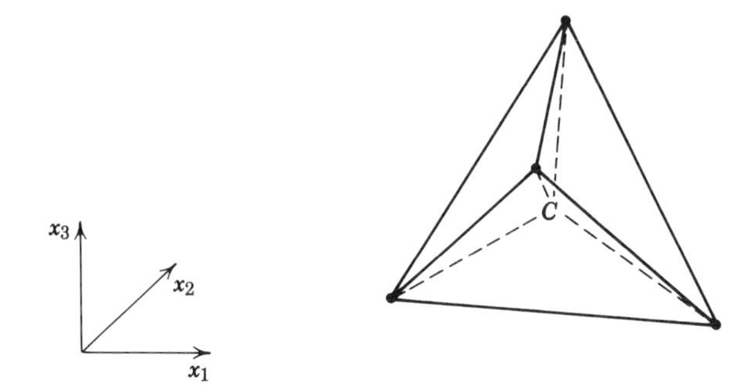

Figure 7.17. A regular simplex design (tetrahedron) in three dimensions showing the equiangular lines from the vertices (design points) to the center C.

stage, we can make a successive progression towards what we hope will be higher and higher response values.

Spendley et al. propose that normally only a single run would be made at each point when completing the next regular simplex and repeat runs would not be made (except insofar as they occur owing to a retracing of the path). It is true that, for the kind of reasons we have described in our discussion of empirical feedback (see Section 7.1), this method will enable us to make progress by empirical feedback. The procedure suffers from the fact that, although it permits empirical feedback, it is not designed to provide scientific feedback. Without replication the reality of any apparent effects would not be satisfactorily established, and it might be dangerous to attempt to use the results from a simplex EVOP for the building of scientific theories designed to lead to a better understanding of the process. Furthermore, even if replication is introduced so that we could be more certain of the reality of whatever effects were seen, the regular simplex design does not allow ready comprehension of just what these effects are. Effects are much more easily understood when the basic design framework is that of a two-level factorial, for example.

It is our opinion therefore that, although this procedure provides an ingenious method of optimization that is probably good for numerical maximization problems, it does not satisfy all of the criteria that we would like to see present in a method of EVOP. In particular, it does not provide the same opportunity of inducing scientific feedback as does standard EVOP.

One claim that has been made for simplex EVOP is that it is, or anyway ought to be, "quicker" than regular EVOP [see, for example, Kenworthy, (1967)]. This idea seems to be based not on theoretical or empirical evidence but merely on a consideration of what would happen in the case where only empirical feedback is of importance and there was no experimental error—a situation which is of little practical interest. It can scarcely be upheld if a less naive view of the working of EVOP is taken and, in particular, if the distinction between empirical and scientific feedback, and the importance of the latter, is borne in mind.

CHAPTER 8

Comments and Questions on EVOP

Experience with the running of EVOP has revealed certain objections, comments, and queries that occur frequently. We take the opportunity to discuss them in this chapter and to recapitulate, in our commentary, some of the principal points already made in previous chapters.

8.1. A DISCUSSION OF SOME OBJECTIONS, COMMENTS, AND QUERIES

My process is too complicated for EVOP.

It is perhaps understandable that each process superintendent usually believes that his is the most difficult and complex process that has ever been operated. He may suggest other processes that he thinks are simpler and are "more suitable" for EVOP. However, experience shows that, when the process superintendents for these other processes are interviewed, they believe, with equal certainty, that *their* processes are too complex for the successful application of EVOP. The best thing to do when a complicated situation exists is to try to learn about it, and this is exactly what EVOP sets out to achieve. Difficult problems do not become less difficult by doing nothing. Possibly the difficulties will be overcome when the effects of the variables are more thoroughly understood. Such understanding can be supplied by an EVOP program.

There are only two important variables on this process. Why look at others?

We know of at least one example where one phase of an EVOP scheme was run for several months with the same two variables being studied in the same design. Literally dozens of cycles were made and no important effects were revealed. The process superintendent believed that the two variables studied were the *only* two of any importance. The superintendent's belief that these variables *were* important was based on good evidence from past experimental work, but the process had, long before, been optimized with respect to these two variables and nothing was gained by their further investigation. When the process and EVOP program were taken over by a more enterprising superintendent, it soon became clear that there were variables which had *not* been previously examined and which were capable of yielding considerable improvement. Very often it is not realized that certain variables are important simply because they have never been changed. Vigilance is needed to ensure that EVOP programs do not bog down. Usually, after a few cycles—say, three, four, or five—if no effects have become apparent, the phase should be abandoned and new variables or new settings of the old variables should be selected. The proper attitude in EVOP is one of impatience. We can, occasionally, lose money by not continuing a phase long enough, but we can certainly lose money by spending too long on a particular phase and thus denying ourselves the possibility of investigating, during the same period, other variables or other settings of the same variables which might have revealed effects of value.

My process is too variable.

In some processes the variation that occurs from one run to another is quite large. The fact of this large variation will sometimes be advanced as a reason for not using EVOP. "We cannot," it is claimed, "obtain sufficiently accurate measurements for EVOP to be useful." As we have seen in Section 1.5, this argument is not valid. If a process has a standard deviation of, say, 10% of the response level, effects of the order of this 10% can be obscured by experimental error and can elude discovery. These same effects can, however, be shown up by performing repeated cycles and averaging results. The existence of large experimental error often indicates a lucrative source of improvement, for it is probable that large effects await detection.

My specifications are too tight to permit EVOP.

Another argument sometimes advanced as a reason for not introducing an EVOP scheme is that, if process conditions are altered at all, the product will no longer meet specifications. Such a claim needs careful investigation. First of all, it is important to determine whether the specifications under discussion are *manufacturing* or *customer* specifications. To encourage tight operation of the process, management often sets manufacturing specifications which are considerably narrower than required by the Sales Department. In such cases alternative control procedures which can be used during EVOP should be negotiable without much difficulty. Customer specifications, on the other hand, represent a somewhat different problem. Although it is nearly always true that written customer specifications are not in complete agreement with existing supplier-customer practice, nevertheless, unless it is absolutely necessary, we would not recommend entering an investigation of the validity of the sales specifications at the time of the institution of an EVOP program. Rather we think it is more sensible to provide a forum in which potential danger to customer relations due to quality variation can be presented. It will almost always develop that adequate safety can be assured with, at most, the introduction of a slight amount of extra testing. In the vast majority of cases, simply an increased level of vigilance on the part of production will provide the necessary safety factor.

Decrease in variability can eventually be obtained only by gaining adequate control of a process. In turn, adequate control requires identification of variables which can cause changes in response. EVOP can achieve this identification of variables as well as estimation of effects. If need be, it is usually better to suffer the temporary inconvenience that might occur in the early steps of this EVOP program because of specification difficulties rather than to attempt to continue operation with inadequate understanding of the process.

As we have already mentioned, a quality control man or someone else concerned with specifications should be present on the EVOP committee. In this way, questions about specifications can be resolved.

Is EVOP the best method of process improvement?

As we explained at the beginning of this book, in the ordinary course of events, many different kinds of endeavors will be going on simultaneously, resulting, in one way or another, in process improvement. Thus, for example, the research department will be developing new basic processes; the experimental and development departments will be carrying out experiments on the small-scale pilot plant, and sometimes on the full-scale plant, to seek process improvements; mathematicians and chemical engineers may be carrying out special computer studies involving process modeling and simulation to try to obtain better understanding of processes, and so on; and interaction among all the various endeavors will also occur. Evolutionary Operation is not intended as a substitute for this work; it is an *additional* means of process improvement. It cannot produce many of the results which are achieved by the more sophisticated techniques. However, it earns its place as a valuable supplement to other process improvement techniques because of its cheapness and potential ubiquity. The specialist techniques must be carried out by skilled technical personnel, by highly trained chemical engineers, chemists, computer scientists, statisticians, and the like. The number of problems to which such personnel can devote their attention at any one time is extremely limited.

How many variables should be included at one time in an evolutionary scheme?

Usually two or three variables can be handled satisfactorily under the normal production conditions with which we are familiar. It may be asked: "Would it not be better to run more sophisticated statistical experiments on the process? It is well known that the efficiency of experimentation increases as the number of variables examined is increased; why, therefore, do we not arrange to run, continuously, experiments in seven or eight variables, employing designs such as complex fractional factorials, instead of working with two or three?" Our experience is that, although sophisticated methods are invaluable for special investigations which are being carefully supervised by special teams of experts, they cannot be successfully carried out on a day-to-day basis, under routine conditions of operation, as a permanent mode of procedure. In particular, it seems next to impossible to run routinely a process in which, say, seven variables are simultaneously being changed. At the planning stage it is often quite difficult to forecast, when seven variables are changed in particular ways, whether or not the process will run satisfactorily at all. In actual operation, the making of changes on too large a number of variables can merely lead to chaos. If we try to introduce such a scheme on the factory floor, it will quickly be abandoned. It is better to have a simple scheme with limited objectives which process personnel can handle, rather than a scheme more theoretically desirable which will never, in fact, be run. It should perhaps be emphasized, once again, that what is being discussed is the normal production situation in which EVOP is applied. For specialist short-period investigations (which, as has been explained, are not the subject of this book but, nevertheless, play an important part in the general scheme of process development to which EVOP is a supplement) the situation is entirely different. In these specialist investigations, in which, for a limited period, it is permissible to interfere with production and in which special supervision and other facilities are made available, techniques of much greater sophistication can be employed. In particular, in a situation where we wished to screen a large number of variables, it would usually be inefficient to consider these in groups of two or three at a time. We would rather examine larger numbers of variables in fractional factorial designs. Thus for example, if 8 factors had to be screened, a 16-run screening experiment might be carried out using a $\frac{1}{16}$ replicate of a 2^8 factorial design [see Box and Hunter (1961)]. Depending on circumstances, we might want to estimate a curved response surface in the region of interest. More complex response-surface designs might then be employed.

In a typical EVOP situation where it is practical to study variables only in groups of two or three at a time, we forego the possibility of detecting dependence (interaction) between variables not in the same group. The effect of this limitation should be minimized as much as possible by examining, in the same group, sets of variables expected to be interrelated. Also periodically, variables whose effects have been found to be important in different phases of operation should be tested together.

What patterns of variants are of most value?

A variety of patterns of variants may be useful for special purposes. Those found most valuable for routine use are the two-level factorial designs often with one or more added points as reference conditions. These simple factorials have these advantages:

1. They are simple to comprehend, perform, and analyze.
2. The inclusion of reference conditions allows continual comparison with a known process and permits the cost of obtaining information to be assessed.
3. Nonlinearity of the surface is easily detected by considering the relative magnitudes of the simple *main effects* on the one hand and *interaction* effects on the other. When the reference conditions are at the center of the design, an over-all measure of curvature can be calculated.
4. Occasionally when elaboration is called for, they can be made the nucleus of more elaborate designs, in particular of composite second-order designs [see Box (1954*b*), Davies (1956), and Cochran and Cox (1957)], by which the surface may be further explored.
5. They lend themselves conveniently to *blocking*, whereby extraneous disturbances due to uncontrolled factors may be reduced (see Section 3.4).

How should past plant records be used in planning the EVOP scheme?

When, as is often the case, past plant records covering long periods of normal operation are available, the planning of an EVOP scheme should always begin with a careful study of these records. In particular, they may be used to determine the approximate magnitude and nature of the uncontrolled variation in the various responses and, consequently, the number of repetitions of the cycle likely to be needed to detect effects of a given size.

From these records we can find out whether the errors in the principal response can be regarded as effectively independent and, if not, we can determine the nature of the dependence that exists. This is of some importance in choosing the period for which each variant should be run before changing to the next variant. In practice, of course, this period also depends partly on convenience of operation and, for continuous processes, on the time it takes for the plant to settle down after a change in reaction conditions.

Suppose, for illustration, we are considering a batch process and wish to answer the question of whether several batches should be made at each set of conditions before changing to another set of conditions or whether the conditions should be changed after each batch. The answer is determined by the nature of the dependence between successive observations. For simplicity, take the situation where we need consider only correlation between adjacent errors, the correlations between errors not adjacent being negligible. If the correlation between successive errors is positive, which will be revealed in the plant records by a tendency for the results to move in smooth trends, then it will be best to change the conditions as often as possible to obtain the most precise estimates of effects. This is the most common situation in practice. However, successive errors are sometimes negatively correlated because of *carryover* effects, where a high-yielding batch tends to be followed by a low-yielding batch, and vice versa. This effect can occur when a certain proportion of a given batch is left in pipes and pumps and is then included in the succeeding batch. Greatest precision in the determination of effects when successive observations are negatively correlated is achieved by running several batches at the same set of conditions before making a change.

In practice, more complicated situations may occur where, for example, there is a tendency for the observations to follow trends at the same time that carryover is present. Also mixing and other phenomena in continuous processes can give rise to even more complex correlation patterns. In a given situation, it may be far from obvious what constitutes the best arrangement of runs, i.e., the arrangement that leads to a minimum variance for the effects.

When past plant data are available and it can be assumed that they form a stationary time series, it is possible to determine the optimal number of runs to be made at each set of conditions by a study of the sample autocorrelations or, equivalently, of the spectrum of the process [see Jenkins and Watts (1968)]. These methods are complicated and beyond the scope of this book. A simple alternative which has been found effective employs EVOP "runs" on past plant data. What we do is to take a period of normal plant operation during which, in fact, no EVOP scheme was being run, and superimpose alternative EVOP patterns onto the observations, observing the natural variation of the "effects" evaluated for each superposition of a pattern.

To illustrate this, consider the data in Table 8.1. This table contains 80 successive observations of percentage yield from a certain process taken during a period when no EVOP program was being run. We might ask the question: "During this period of operation, would it have been better to have conducted a two-variable EVOP scheme by running the usual five variants in the pattern 1, 2, 3, 4, 5; 1, 2, 3, 4, 5; . . ., 1, 2, 3, 4, 5; . . ., which we shall call pattern 1, or would it have been better to run 1, 1, 2, 2, 3, 3, 4, 4, 5, 5, . . . (pattern 2), or 1, 1, 1, 2, 2, 2, 3, 3, 3, 4, 4, 4, 5, 5, 5, . . . (pattern 3), or 1, 1, 1, 1, 2, 2, 2, 2, 3, 3, 3, 3, 4, 4, 4, 4, 5, 5, 5, 5, . . . (pattern 4), or what?" We can superimpose the foregoing patterns on the 80 past observations to obtain 1, 2, 3, 4, 5 sixteen times 1, 1, 2, 2, 3, 3, 4, 4, 5, 5 eight times, and so on.

Table 8.1. Eighty successive observations of percentage yield

1	61.21	21	64.51	41	62.69	61	61.48
2	63.09	22	65.35	42	62.82	62	61.83
3	61.31	23	65.07	43	63.66	63	64.28
4	61.46	24	63.51	44	60.93	64	63.81
5	63.09	25	64.46	45	61.26	65	63.63
6	62.78	26	65.85	46	62.77	66	62.84
7	63.95	27	64.96	47	62.64	67	64.78
8	63.44	28	64.53	48	63.28	68	64.27
9	64.24	29	64.40	49	61.54	69	64.89
10	62.78	30	62.96	50	62.13	70	65.63
11	64.92	31	62.79	51	60.44	71	64.78
12	64.14	32	64.26	52	61.94	72	64.90
13	62.87	33	64.89	53	62.77	73	66.55
14	62.34	34	62.51	54	64.05	74	63.50
15	65.39	35	63.14	55	62.88	75	65.98
16	62.34	36	63.55	56	63.02	76	64.30
17	64.52	37	63.26	57	63.24	77	67.29
18	65.04	38	64.12	58	62.60	78	64.83
19	63.51	39	62.58	59	61.20	79	66.53
20	63.51	40	62.56	60	64.62	80	65.01

Imposing the pattern 1, 2, 3, 4, 5; 1, 2, 3, 4, 5; etc., we obtain 16 estimates of the A effect using the formula

$$A = \tfrac{1}{2}(y_3 + y_4 - y_2 - y_5).$$

These 16 estimates provide an estimate

$$V_1(A) = \sum_{i=1}^{16} \frac{(A_i - \bar{A})^2}{15} = \frac{15.473}{15} = 1.03,$$

where A_i ($i = 1, 2, \ldots, 16$) are the 16 A estimates and $\bar{A}$ is their average. Similar calculations can be made for the B effect [$V_1(B) = 0.95$], the AB interaction [$V_1(AB) = 0.94$], and the change in mean (CIM) effect [$V_1(\text{CIM}) = 0.92$].

To provide a basis of comparison for the different schemes, we can now ask how accurately would the mean A effect have been measured by using the whole 80 observations with the different patterns. To do this, we use the fact that, although individual observations are correlated, estimates from successive cycles will be very nearly uncorrelated. For example, with pattern 1 we should complete 16 cycles so that $V_1(A)/16$ is an estimate of the variance of the A effect from the *whole* experiment, using this pattern. For the four patterns mentioned previously, we should obtain the following results:

Variance of A for Whole Set of Runs	Pattern
$V_1(A)/16$	1
$V_2(A)/8$	2
$V_3(A)/(5\frac{1}{3})$ [1]	3
$V_4(A)/4$	4

[1] The last five observations are omitted to form five complete sets of 15 observations with pattern 3.

We could obtain similar estimates for the main effect B, the interaction AB and the change in mean effect (CIM). We can now combine these results in a weighted sum. Thus for pattern 1, we can take *

$$Q_1 = \tfrac{1}{16}[V_1(A) + V_1(B) + V_1(AB) + \tfrac{5}{4}V_1(\text{CIM})]$$

as a measure of the intrinsic variance resulting from the use of pattern 1, and this will provide an estimate of the average variance for a main effect or interaction from the complete set of 80 observations used. Similar weighted averages of $V_2(A)/8$, etc., are taken for pattern 2, and so on. For the particular set of data given in Table 8.1 we obtain the following results:

Pattern	Estimated Variance of Main Effect or Interaction for 80 Runs
1	0.255
2	0.154
3	0.195
4	0.171

*Q_1 can alternately and equivalently be calculated by averaging the within-cycle sample variances. Thus, if for the ith cycle of pattern 2 we denote the averages at the five sets of conditions by $\bar{y}_{1i}, \bar{y}_{2i}, \bar{y}_{3i}, \bar{y}_{4i}$, and $\bar{y}_{5i}$, then

$$Q_2 = \frac{1}{8}\sum_{i=1}^{8}\sum_{j=1}^{5}(\bar{y}_{ji} - \bar{y}_i)^2,$$

where $\bar{y}_i = \sum_{j=1}^{5} \bar{y}_{ji}/5$. (Similar formulas apply whenever the observations can be divided into an exact number of cycles with no observations omitted.)

In this particular example, it was known that carryover occurred, and it might therefore have been expected that patterns involving successive repetition of conditions would have been preferred. This, in fact, turned out to be the case, pattern 2 being the best choice. The trends which also occurred in the data evidently ensure that the maintaining of constant conditions for *more* than two batches is somewhat less desirable.

Where we have parallel units, is it necessary to run all the parallel units on EVOP simultaneously?

No. When we can rely on the essential similarity of the individual units, we need perform EVOP on only a single unit which can act as a *bird-dog* or *indicator unit* for all the others. In particular, it should be noticed that, where we have, say, eight parallel units, only one of which is being "EVOP-ed," then, if the final product from the unit can be mixed or blended off with the product from the other seven units, we can afford to be rather more daring than usual in our choice of the variants to be run. We do, of course, have to be certain that separate samples can be taken from the bird-dog unit. In some instances special instrumentation, sampling facilities, etc., have been introduced on the bird-dog unit. In the building of new plants, the deliberate introduction of facilities of this kind *ought to be considered as a routine matter.*

Should the variants be run in random order?

Faced with the possibility of serial correlation between successive observations, the statistician would normally wish to perform the variants in random order within each cycle, thus guaranteeing the validity of the standard analysis. However, it is usually much simpler to run a systematic routine of variants on the plant than a random one, particularly when the time for running each cycle is short. In practice a standard run order is usually adopted, as illustrated by the examples in Chapters 4 and 5. The effect of not randomizing in this particular context will usually be small. To see this consider the observations after n cycles of k variants written down in a table having n rows and k columns. We notice that we are only concerned with comparisons of *column* means. Now the major part of the dependence occurs *within* rows, and in this situation, as was shown by Box (1954*a*), the standard analysis is usually not seriously invalidated.

How should multiple responses be considered?

Although it is possible in theory to combine all responses into a single criterion, such as profitability, this usually presents great practical difficulties. As a general rule, it is best to represent the problem as one of improving a *principal* response (for example, the observed cost of manufacturing a pound of product at the relevant process conditions), which is often some function of two or more observed quantities, subject to satisfying certain conditions on a number of *auxiliary* responses. These auxiliary responses usually measure the quality and important physical properties of the product.

Very careful thought in the selection of the principal response is essential. The vital question to ask is: "If this response is improved, will it mean necessarily that the process is improved?"

In the examples we have discussed, the reconciliation of the requirements for the various responses in the light of the experimental results was done by looking at approximate contour diagrams. This approach has the virtue of simplicity and is fairly satisfactory in the situations specifically dealt with here when there are only two or three variables to consider. It has been pointed out in Section 7.3 that the problem is exactly like that of linear programming, with the added complications that the problems are not always approximately linear and that the restraints are not known exactly but must be estimated. Consequently, the use of formal programming methods could scarcely be justified in the typical EVOP situation.

How best can the stream of information coming from the plant during the evolutionary process be presented to those responsible for deciding what to do?

Two things are necessary: first, to show how much weight ought to be attached to the results and, second, to present them in such a way that their interpretation is facilitated as much as possible.

To convey a sense of the degree of reliability which the plant manager should associate with the results, several ideas have been tried. In the original schemes various types of sequential charts and significance tests were used. It is now felt, however, that the problem is not one of significance testing, and what is needed is a presentation of the information contained in the data, unweighted by external features subsequently injected into the situation; for example, the particular choice of the risks α and β and of the hypotheses "which it is desired to test" (subtleties not readily comprehended by plant personnel) can completely alter the apparent implications of a set of data when these are plotted on a sequential chart. If the observations are roughly normally distributed, are independent, and have constant variance, then all the information they contain is included in the averages, the standard error, and the number of observations. These statistics seem best comprehended in the form of effects with 2 S. E. limits. In appraising the results, other information about the importance of different sorts of effects must be used. It seems best, however, to separate this from the presentation of the results, which then refer to information supplied by the observations and to nothing else.

To show the implications of the average results, once it becomes clear that these are determined sufficiently accurately, there is no doubt that for two or three variables geometrical representation is ideal. It allows the general trend in the responses and their relationship to each other to be appreciated in a manner not possible in any other way.

The process superintendent should run EVOP in much the same way as he would play a card game. The information board shows him his "hand" at any given time, and depending on that hand there are a number of actions he can take (including drawing a further card and deciding what to do then).

When one or more of the variants is clearly better than the works process or when clear-cut trends in the results exist, the process superintendent will have no difficulty in following the indications of the information board. Where the results indicate that complexity exists, he will be able to obtain the help of the statistician on the EVOP committee in elucidating the results and, where necessary, in augmenting and modifying the cycle of variants in the next phase of operation so as to resolve the complexity.

If the full information board is kept where process workers can readily see it, this provides added interest and is an incentive to accurate operation, which itself can result in a general improvement in productivity. In some instances, it may be more convenient for a separate display to be maintained in a place where it can be seen by the process workers. Such a display could omit the more intricate details of the effects of the variables, and so on, and might look like that shown in Section 1.6.

In what way can the results from small-scale experimental studies be used in planning the evolutionary scheme, and how should this affect the way in which these small studies are conducted?

The complexity mentioned in the foregoing arises principally because the variables studied fail to behave independently in their effects on the response; that is, they interact. The plant process will usually have been arrived at as the result of a small-scale investigation of at least some of the variables. This investigation should have culminated in a study of the local "geography" of the response surfaces in the neighborhood of the proposed operating conditions. The principal features of the laboratory response surfaces will normally be preserved on the plant scale even though some distortion occurs. If the laboratory investigation has been conducted with carefully designed experiments and, in particular, with response-surface designs, it will frequently be possible to discover transformations of the variables originally considered which act approximately independently, at least for the principal response. Thus, for example, in a chemical context we may find that it is better to think in terms of *ratios* of certain concentrations rather than of the individual concentrations. By working in terms of these new variables difficulties due to complexity of the surface can be greatly reduced in the plant-scale investigation. Unfortunately it will rarely be the case that after the initial stages of an EVOP program the variables being considered will be those that have been studied in the small-scale work. Thus small scale studies, while helpful, usually have only a limited value.

In practice it is impossible to attain truly static operation. Small variations in the process conditions are bound to occur from one run to the next. In cases in which these changes are recorded why should we bother to carry out a special pattern of variants? Why not use the "pattern of variants" supplied by the natural variation of the process to supply information on which evolutionary improvement can be based?

Suppose the level of a response is dependent on k variables and that, owing to imperfect control, deviations $x_1, x_2, \ldots, x_k$ occur from the settings of the variables called for by the standard process resulting in a corresponding deviation y in the response. Suppose finally that, to a sufficient approximation, the deviation y is linearly related (apart from an error e) with the deviations $x_1, x_2, \ldots, x_k$, so that

$$y = \beta_1 x_1 + \beta_2 x_2 + \cdots + \beta_k x_k + e.$$

Then it might be thought that we could use the naturally occurring values of $y, x_1, x_2, \ldots, x_k$, as data to obtain least-squares (that is, multiple regression) estimates of $\beta_1, \beta_2, \ldots, \beta_k$ and so to measure the individual effects of the variables. Knowledge of these effects would, we might expect, show us how to improve the process. Although in the past the use of this method would have been laborious, now, with the use of the electronic computer, the necessary calculations could be made in seconds. At first sight such a method appears attractive, for here we seem to have an evolutionary scheme in which we do not need to bother about introducing variants deliberately. On closer examination, however, its value proves to be much more doubtful. Many investigations have been made by statisticians in which plant records have been analyzed by multiple regression in an attempt to determine the "effects" of the variables and so to improve the process. In our experience the results of such investigations have nearly always been disappointing. The reasons are not hard to find:

1. In order to determine the effect of changing the level of a variable,

we must actually change it. It is not enough to observe the variable. The reason is that, even if in naturally occurring data a variable x_i has a statistically significant coefficient b_i, the variable x_i may not actually *cause* the changes observed in the response y but may simply be correlated with another variable (perhaps an unobserved, latent, or "lurking" variable) that *does* cause the change in response.

2. Many of the factors that may vitally affect the process are not, in the normal course of events, altered at all.

3. Those factors that vary naturally do so, not over the ranges we should like but over ranges dictated by the degree of control which happens to exist. The more control is improved, the less information we get. Thus the variables that are most rigidly controlled, although likely to be those which would produce the largest effects if allowed to vary, may produce no detectable effects.

4. The fluctuations that occur naturally in the variables are often heavily correlated. This is because in normal operation variables tend to be changed together. This results in poor precision of the estimates and presents difficulties in disentangling the effects of the variables.

5. A particular aspect of the point discussed under (1) concerns the spurious effects associated with time. Accidental modifications often tend to happen in "phases" and so become spuriously correlated with causally unrelated happenings and trends in time. Such effects can lead to completely wrong conclusions. Attempts to eliminate time trends computationally usually eliminate the effects of the factors at the same time.

What this all amounts to is that a naturally occurring scheme of variants is not very likely to provide a good, or even passable, "design" and, consequently, the amount of information generated by natural variation may be scarcely worth salvaging.

A fuller discussion of the first point is given by Box (1966*b*) and is summarized in the following.

Suppose that in a chemical process it has been found that undesirable frothing can be reduced by increasing pressure. The standard operating procedure is, therefore, to increase pressure whenever frothing appears. Suppose that the frothing in fact occurs because of an unsuspected impurity (which is, of course, not measured because it is unknown). Suppose, finally, that a high value of this impurity not only produces frothing but also lowers yield, but that yield is unaffected directly by a change in pressure. Now let y denote the deviation from average yield, x_1 the deviation from average pressure, and x_2 the deviation from average impurity level. Then, if a "regression of yield on pressure" $\hat{y} = b_1 x_1$ is fitted by the usual least-squares procedure [Draper and Smith (1966)], we may well find a highly significant coefficient b_1.

This well-known phenomenon of "nonsense" correlation exhibited in this example is worth studying further. Suppose that in the relationship

$$y = \beta_1 x_1 + e$$

the error e is entirely produced by the deviation x_2 in the impurity so that $e = \beta_2 x_2$. Then, in the example, there would in reality be an exact relationship $y = \beta_1 x_1 + \beta_2 x_2$ connecting y with the two deviations x_1 and x_2 (with $\beta_1 = 0!$). Now, of course, the actual levels of x_2 are unknown, but suppose $\hat{x}_2 = a x_1$ is the formal regression of x_2 on x_1, which would be obtained if the values of x_2 *were* available. Then it is readily shown that

$$b_1 = \beta_1 + a\beta_2. \tag{8.1.1}$$

In this expression β_1 is zero and we appear to obtain a real effect only because of the influence of the bias term $(a\beta_2)$. On the other hand, by using (8.1.1) we see that our fitted equation $\hat{y} = b_1 x_1$ which ignored x_2 can be written $\hat{y} = \beta_1 x_1 + \beta_2 a x_1$ or as

$$\hat{y} = \beta_1 x_1 + \beta_2 \hat{x}_2. \tag{8.1.2}$$

Provided the system *continues to be run in the same fashion as when the data were recorded,* we can use pressure to indicate the level of y. On the other hand, the value of b_1 will be utterly misleading if interpreted as the effect on the variable y of a unit *change* in x_1. If we hope to increase yield by reducing pressure, we will be disappointed. A similar argument applies for any number of variables. [See, also, Box and Coutie (1956, p. 111).]

Can EVOP be made automatic?

The procedure of EVOP so far described is a "manual" one. It requires no special facilities and can be immediately applied in one form or another to a very large proportion of industrial processes. This is so whether the available plant is of the crudest kind or whether it includes such refinements as automatic controllers and recorders. The process superintendent is himself a part of the "closed loop," thus ensuring that sensible action will be taken even in unforeseen circumstances.

With a sufficiently instrumented plant, the empirical feedback part of the EVOP procedure is, of course, capable of being made completely automatic. Thus variables whose levels are regulated by a controller can be automatically changed at regular intervals so as to follow a cycle of variants, and a response such as cost per pound can be automatically computed from the readings of instruments which measure the properties of the product. The cumulated differences in response at the various process conditions can be used to trigger off adjustments in the location of the pattern of variants, so completing the evolutionary process.

In continuous processes (in which there is a continuous input of starting materials and a continuous output of product) the "pattern of variants" can consist of a continuous locus instead of being a discrete set of points. In fact, we can introduce a sinusoidal probing signal or a set of such signals at different frequencies. The problem of detecting the effects of the variables is then precisely that which arises in communication theory, namely, of detecting a signal of known form in a noisy channel. As we have explained in Section 7.3, it is best to think of systems of this kind as processes of automatic optimization.

The introduction of automatic optimization would usually be worthwhile only if the response surface itself were changing in some way and it is desirable to attempt to follow that change. For most chemical processes the response surfaces are reasonably stable. In some, however, unpredictable but steady changes can occur owing to slow composition changes in raw material (such as crude oil) or in catalyst activity. Here unpredictable differences in the position of the optimum conditions may occur between batches of catalyst and also within the life of a single catalyst batch. In these cases automatic optimization may be effective in keeping the plant operating near its best performance, but only if the rate at which information is generated is sufficiently large compared with the rate at which the optimum conditions are changing.

As we have explained in Section 7.3, although EVOP and automatic optimization do possess points in common, the philosophy of the two procedures is quite different. In the case of EVOP we have a procedure that needs no special instrumentation or facilities and that explores and screens a large number of variables of every kind. In the case of automatic optimization we are dealing with some specific part of the process which, it is known, needs to be continuously optimized, and this is being done by means of very sophisticated equipment. For further reading on automatic optimization see Kotnour et al. (1966) and Box and Chanmugam (1962).

Where can more be learned of actual industrial experience with EVOP?

An excellent review of industrial applications of EVOP has been presented by W. G. Hunter and J. R. Kittrell (1966). This review, which lists 68 references, discusses applications of EVOP in a wide variety of environments. Among these, applications in the chemical industry are most numerous, with special references to uses by the Dow Chemical Company [Anonymous (1961*a*, 1961*b*)], American Cyanamid Company [Cestoni et al. (1960); Koehler (1958*a*, 1959*a*, 1959*b*)], Imperial Chemical Industries Limited [Coutie (1959*a*, 1959*b*)], Tennessee Eastman Company [DeBusk (1962); Pursglove (1961); Wilson (1960)], the Chemstrand Corporation [Annual Report (1959); Riordan (1958, 1959*b*)], Phillips Petroleum Company [Pursglove (1961); Weaver (1963), Standard Oil of Ohio [Klingel and McIntyre (1962); Pappas (1962)], and Monsanto Company [Anonymous (1960); Hehner (1963*c*)]. Several applications in the food industry are also mentioned, in particular use by Swift and Company [Samuel (1962)], Canadian Packers Limited [Samuel (1962)], and the A. E. Staley Manufacturing Company [Koleff (1963)]. Also discussed are uses of EVOP by the canning industry [Filice (1963)], and by the Maumee Chemical Company in such diverse projects as saccharin production, a biocide for sea lampreys, isatoic anhydride, anthranilic acid and benzotriazole processes [Anonymous (1963)]. Other applications are mentioned in the automotive industry where EVOP was applied to resistance welding of automotive sheet metal [Thomas and Webster (1960)] and other manufacturing operations [Thomas (1965)]. Among these applications a savings of several hundred thousand dollars per year by the Chemstrand Corporation has been mentioned by F. S. Riordan (1958, 1959*b*), and a savings of $250,000 per year for two products has been reported by H. O. Hehner (1963*c*) in the Monsanto Company. In the description of uses by Tennessee Eastman Company, reference is made to a report by R. E. DeBusk (1962) that EVOP was used in approximately 15 processes with a savings of from $15,000 to $50,000 per year per process.

APPENDIX 1

The Approximate Method of Estimating the Standard Deviation in *EVOP**

Suppose there are k operating conditions and n cycles have been performed. The following data will then be available:

	Condition Number				
	1	2	3	. . .	k
Overages after $n - 1$ cycles	$\bar{y}_{1,n-1}$	$\bar{y}_{2,n-1}$	$\bar{y}_{3,n-1}$	. . .	$\bar{y}_{k,n-1}$
Dbservations in nth cycle	y_{1n}	y_{2n}	y_{3n}	. . .	y_{kn}
Aifferences $d_{in} = y_{in} - \bar{y}_{i,n-1}$	d_{1n}	d_{2n}	d_{3n}	. . .	d_{kn}

Each difference [as calculated on line (iv) of the EVOP form] is the difference between an individual observation (whose variance is σ^2) and an independently distributed average of $n - 1$ observations (each with variance σ^2). Consequently the variance of each of the differences is

$$\sigma^2(d_n) = \sigma^2 + \frac{\sigma^2}{n-1} = \frac{n\sigma^2}{n-1},$$

so that

$$\sigma = \sigma(d_n)\left(\frac{n-1}{n}\right)^{1/2}.$$

For simplicity we use an estimate of $\sigma(d_n)$ based on the range R_n of d_{1n}, d_{2n}, . . . , d_{kn}. As we have already noted in Section 2.5, an estimate of the standard deviation of a normal distribution can be obtained by multiplying the

* See Section 4.2.

sample range by a factor w given in Table 2.3. A more extensive tabulation is given in Table IV. An estimate of $\sigma(d_n)$ is provided therefore by $R_n w_k$, and on multiplying this quantity by $\sqrt{(n-1)/n}$ we obtain an estimate s_n of σ contributed by the nth cycle. We have

$$s_n = R_n \left(\frac{n-1}{n}\right)^{1/2} w_k,$$

that is,

$$s_n = R_n f_{k,n},$$

where

$$f_{k,n} = \left(\frac{n-1}{n}\right)^{1/2} w_k$$

is the factor evaluated and recorded in Table V and also given on the EVOP forms; for example, when $k = 5$, $n = 2$, $w_k = 0.4299$, and $f_{k,n} = 0.4299/\sqrt{2} = 0.30$ as in the table.

After each cycle of the EVOP scheme except the first, an estimate of σ can be calculated in this way. After n cycles, we have $n - 1$ estimates, $s_2, s_3, \ldots, s_n$, and the average $(s_2 + s_3 + \cdots + s_n)/(n-1)$ provides the current estimate of σ.

We have included, on the EVOP forms, a table of useful factors for use in the calculations. For convenience these factors are also given in Table VI.

APPENDIX 2

Generating Data for the *EVOP* Game*

We have seen in Section 6.2 how we can become familiar with the calculation procedures of EVOP by using data "generated" in the classroom rather than on the plant. The data are generated by randomly drawing counters from a number of bowls, each of which represents a particular set of conditions. On each counter is written the value of one or more responses. We here consider how the counters for such a game may be prepared.

We can imagine contours for some response drawn across the points of a two-variable EVOP scheme, as shown in Figure A2.1. These contours represent the "true" response surface. Realistic contour maps for two or three responses that are to be considered should be prepared. Where possible it is a good idea to base the game on actual experience with familiar variables and to have a situation unfolding which is of interest to the particular audience.

Suppose we wish to simulate a situation in which the variance $V(y)$ of the response under study is $\sigma^2 = 4$. Consider, for example, the preparation of the counters that will represent observations made at the center conditions of the design. Interpolating in Figure A2.1, we see that the true response at this set of conditions is 56.6. We then need to obtain a sample of normally distributed observations with mean 56.6 and variance 4. We denote this population by $N(56.6, 4)$. The counters that will fill the bowls representing the various conditions should therefore be

Point 0: observations from $N(56.6, 4)$.
Point 1: observations from $N(55.6, 4)$.

* See Section 6.2.

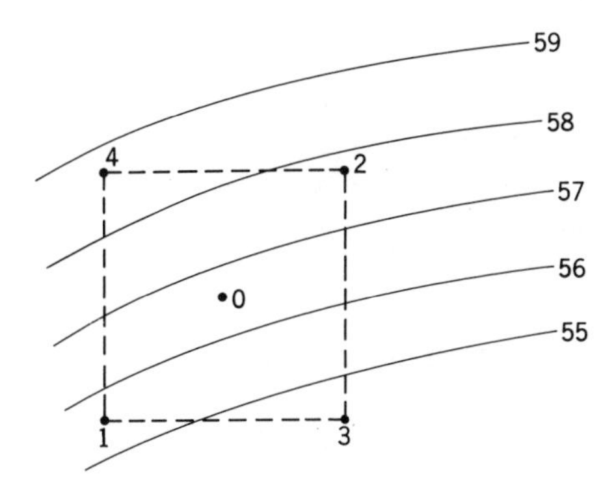

Figure A2.1. *"True" response contours for a particular sequence in the EVOP game.*

Point 2: observations from $N(57.8, 4)$.
Point 3: observations from $N(54.5, 4)$.
Point 4: observations from $N(58.7, 4)$.

Point 0 data is then generated as follows. A table of random normal deviates such as Table VII, for example, provides "observations" from a $N(0, 1)$ distribution. We proceed by taking a moderate sized sample of these observations (about 40 to 50 members), multiplying each observation by $\sigma = 2$, and adding the result to the mean $\mu = 56.6$. This provides 40 to 50 observations from the normal population $N(56.6, 4)$. [In general, if z is drawn from a $N(0, 1)$ distribution, then $z' = \mu + z\sigma$ will have a $N(\mu, \sigma^2)$ distribution.] A similar procedure using different samples from the table of normal deviates will provide samples of observations for points 1, 2, 3, and 4, which are then recorded on the cardboard counters.

The five sets of counters are now placed in five bowls appropriately labeled. The first cycle of the two-variable EVOP scheme is "run" by selecting a counter from each of the five bowls and reading off the results. After the calculations for the first cycle have been made and the counters have been returned to the bowls and remixed, observations for the second cycle can be drawn, and so on. In practice several responses will usually be studied simultaneously and a complete set of response values can be recorded on each counter as is done in the EVOP game in Section 6.2.

APPENDIX 3

Optimal Empirical Feedback*

The reason that it is important to distinguish between empirical and scientific feedback is that the strategy of experimentation which should be adopted depends on the type of feedback to be induced. It can be shown, no matter what the size of the experimental error and over a wide range of circumstances, that to provide optimal empirical feedback, modifications should be judged from the result of a single trial. One should take no account of the possibility that an observed "effect" might be due to chance. On the other hand, to provide effective scientific feedback, it is necessary that each set of conditions be sufficiently repeated so that the effects on which the technologists' decisions are based are established with some certainty.

The argument that optimal empirical feedback can, in some circumstances, be achieved by acting on the result of a single trial, whatever the standard deviation, can be arrived at from a number of different starting points. We here present two very different arguments (Box, 1966*a*) which lead to this same conclusion. It is admitted, of course, that in both cases the model which is envisaged is a very simplified one. These arguments have been briefly outlined in the text (See Section 7.1) and are here presented in more detail.

A3.1. CHOICE OF STRATEGY USING EMPIRICAL FEEDBACK TO ACHIEVE THE GREATEST GAIN IN A FIXED TIME

Introduction

Suppose that during the routine operation of an industrial process a series of modifications is tested, one at a time. Any modification which

* See Section 7.1.

yields satisfactory test results is immediately included in the process while any modification failing the test is discarded and not retested. Suppose we are to continue the procedure for a total period L, which may be the whole life span of the process. We can, in this available time L, decide to test a *large* number of modifications but, since each one can then only be tested for a short period, we shall run a high risk of "wrongly" including or discarding each one. We can, on the other hand, reduce these risks by running *fewer* modifications, each for a longer time. This can only be done, however, at the expense of *not exploring the possibilities of a wider population of modifications.* How should we proceed so as to make the most profit? Some light can be thrown on this question by considering the consequences of the following somewhat idealized mathematical model.

Mathematical Model

We define the *effect* u of a process modification as the increased revenue per unit of time resulting from its inclusion, and we assume that the effects of different modifications are additive. We suppose that all the modifications that might conceivably be tried form an *effect distribution* with mean m. For convenience the monetary unit is chosen so that the effect distribution has unit standard deviation.

The selection of a modification for test is then simulated by making a random drawing from this distribution. After a particular modification has been selected, it is tested by making an actual trial on the process. The details of the conduct of the trial are not essential to this derivation, but typically the process would be run with and without the modification, the average difference in performance providing an estimate $\bar{x}$ of its effect. In any case we denote by n the *time period* for which the plant is occupied in making the trial, and we assume that the variance $\sigma^2(\bar{x})$ of the estimate of the effect is inversely proportional to n so that $\sigma^2(\bar{x}) = \sigma^2/n$. For an industrial process with a continuous throughput this idealized model would be appropriate if the noise followed a Wiener process. In practice this could never occur exactly; however, the model would be adequate provided the time period n devoted to the test of a particular modification was not taken so small that serial correlation effects became important. For an industrial process in which the product was manufactured as a series of equally sized batches, $n = 1$ could correspond to the shortest trial possible (usually the comparison of one batch made with the modification and one batch made without it). Longer trials would consist of an integral number of these basic comparisons. If the test results were distributed independently, then, as before, $\sigma^2(\bar{x}) = \sigma^2/n$, in which σ^2 would now be the variance of the basic comparison. The results could then be applied to a batch process for integral values of n. We suppose that the standard test procedure is to include the modification if, after a test period of size n, $\bar{x}$ is larger than some value ξ, and otherwise to discard it. Our problem is to choose ξ and n so as to yield maximum total profit.

It seems reasonable to expect that the effect distribution will be unimodal, very good or very bad modifications both being rare occurrences, and we shall assume this distribution to be normal. We shall use the notation

$$f(z) = (2\pi)^{-\frac{1}{2}}e^{-z^2/2}, \qquad F(\zeta) = \int_{\zeta}^{\infty} f(z)\, d(z), \qquad \text{(A3.1.1)}$$

for the normal probability density and the normal probability integral.

If we decide to test each modification for a period n, there will be $L/n = p$ test periods available. The number p is not necessarily an integer, and we denote by $[p]$ the next integer below p. This is, therefore, the actual number of modifications that we will be able to test in the time available. With a modification having a true effect u we suppose $\bar{x}$ will have density function

$$f\left[\frac{n^{\frac{1}{2}}(\bar{x} - u)}{\sigma}\right],$$

and the chance of accepting such a modification is

$$F\left[\frac{n^{\frac{1}{2}}(\xi - u)}{\sigma}\right].$$

Now for fixed m the average profit or loss accumulated *during the test period itself* will be fixed since it depends only on the effect distribution $f(u - m)$. For our present purpose, therefore, we ignore this contribution.

Results

If a particular modification is included as a result of testing it in the ith test period, a profit or loss is made proportional to $uL(p - i)/p$ in the time $L(p - i)/p$ remaining in which the modification may be usefully applied. The total expected profit to be made during the whole period L is, therefore,

$$T = \frac{L}{p}\sum_{i=1}^{[p]} (p - i)\int_{-\infty}^{+\infty} uf(u - m)F\left[\frac{n^{\frac{1}{2}}(\xi - u)}{\sigma}\right] du. \qquad \text{(A3.1.2)}$$

Now

$$\sum_{i=1}^{[p]} (p - i) = \tfrac{1}{2}[p](2p - [p] - 1) = \tfrac{1}{2}p(p - 1) + \tfrac{1}{2}\delta(1 - \delta), \qquad \text{(A3.1.3)}$$

where $\delta = p - [p]$. The maximum value of $\frac{1}{2}\delta(1 - \delta)$ is $\frac{1}{8}$, occurring when $\delta = \frac{1}{2}$. This contribution can usually be ignored in comparison with $\frac{1}{2}p(p - 1)$, and the left-hand side of (A3.1.3) will henceforth be approximated by

$$\tfrac{1}{2}p(p-1) = \frac{1}{2}\,\frac{L}{n}\left(\frac{L}{n} - 1\right).$$

This, of course, provides the exact value whenever $L/n = p$ is an integer. If we apply this approximation and if, in addition, we evaluate the integral in (A3.1.2) using, for example, methods described by Grundy et al. (1956), we obtain

$$T = \frac{1}{2}\,L\left(\frac{L}{n} - 1\right)s^{-3}\int_{\xi+m(s^2-1)}^{\infty} zf\left(\frac{z - ms^2}{s}\right)dz, \qquad \text{(A3.1.4)}$$

where $s^2 = 1 + \sigma^2/n$. Equivalently, we can write

$$T = \frac{1}{2}\,L\left(\frac{L}{n} - 1\right)\left[mF\left(\frac{\xi - m}{s}\right) + s^{-1}f\left(\frac{\xi - m}{s}\right)\right]. \qquad \text{(A3.1.5)}$$

Conditions for a Profit. From (A3.1.5) profit, breakeven, or loss is achieved as

$$\frac{f[(\xi - m)/s]}{F[(\xi - m)/s]} + ms, \qquad \text{(A3.1.6)}$$

is positive, zero, or negative. Thus, in particular, if m is nonnegative, a profit will always be made, however ξ and n are chosen. With m negative, however, a loss can be made for some values of ξ and n.

Optimal Value of ξ. From (A3.1.4), for any n the optimal value of ξ is

$$\xi_0 = \frac{-m\sigma^2}{n}, \qquad \text{(A3.1.7)}$$

and the profit obtainable using this value is

$$T_{\xi_0} = \frac{1}{2}\,L\left(\frac{L}{n} - 1\right)[mF(-ms) + s^{-1}f(-ms)]. \qquad \text{(A3.1.8)}$$

Optimal Value of n when ξ ***is Optimally Chosen.*** Differentiating with respect to n, we obtain

$$\frac{\partial T_{\xi_0}}{\partial n} = \frac{-L^2}{4n^2 s}\left[2msF(-ms) + \left(1 + \frac{\sigma^2/L + 1}{s^2}\right)f(-ms)\right]. \qquad \text{(A3.1.9)}$$

When m is positive or zero, $\partial T_{\xi_0}/dn$ is negative for all values of n. T_{ξ_0} is then a monotonic decreasing function of n, so that whatever the (positive) value of m, T_{ξ_0} is maximized when n is made as small as possible.

When m is negative, the optimal value of n may be found by setting (A3.1.9) equal to zero, that is, by solving the equation

$$-2ms\frac{F(-ms)}{f(-ms)} = \frac{L(s^2+1)+\sigma^2}{Ls^2}. \qquad \text{(A3.1.10)}$$

Optimal n for General ξ. It is also of interest to consider the optimal value of n for general ξ. We have

$$\frac{\partial T}{\partial n} = \frac{-L^2}{4n^2s}\left\{2msF\left(\frac{\xi-m}{s}\right)\right.$$
$$\left. + \left[1 + \frac{\sigma^2/L+1}{s^2} + \sigma^2\left(\frac{1}{n}-\frac{1}{L}\right)Q(\xi)\right]f\left(\frac{\xi-m}{s}\right)\right\}, \qquad \text{(A3.1.11)}$$

where

$$Q(\xi) = \frac{\xi-m}{s^2}\left(\frac{\xi-m}{s^2}+m\right).$$

Now $Q(\xi)$ is nonnegative for $\xi \leqq -m\sigma^2/n$ or $\xi \geqq m$. Therefore

1. When m is zero, T is maximized *for all* ξ when n is made as small as possible.
2. When m is positive, for any fixed ξ satisfying the inequality $-m\sigma^2/n < \xi < m$, the optimal value of n depends on m and σ, and may be found by equating (A3.1.11) to zero. For any other value of ξ, T is maximized when n is taken as small as possible *whatever* the actual values of m and σ.
3. When m is negative, for any fixed value of ξ, the optimal value of n depends on m and σ and can be found by equating (A3.1.11) to zero.

Discussion

The conclusions are seen to depend on the mean of the effect distribution. This parameter determines the number of good modifications among those tested. When m is zero, we have a situation where there are as many good modifications as bad ones. When m is negative, there are more bad than good modifications, and when m is positive, more good than bad ones. Discussions with industrial experimenters have brought conflicting opinions as to the most likely situation. In favor of a positive m it has been argued that there would be more good modifications than bad ones because the population of modifications under consideration is limited to those proposed as *worth trying*. Conversely, it has been argued that an operating process should be already close to optimum conditions and, therefore, most modifications will make it worse. This counterargument would seem to apply only to modifications which already had been carefully tried and not to entirely new ideas. It seems impossible to arrive at any general rule, and it is convenient to begin our discussion by considering first the case $m = 0$ and then seeing how far the conclusions appropriate to that case are modified for other choices of m. When $m = 0$ we are led to the following, somewhat surprising, conclusions:

1. The largest possible number of modifications should be tested (with the smallest possible test period) whatever σ and whatever ξ.
2. The best value of ξ is zero.

With $\xi = 0$ and with n_0 the smallest possible test period, the average profit is proportional to

$$T_0 = \frac{(L/n_0)^{1/2} - (L/n_0)^{-1/2}}{(n_0/L + \sigma^2/L)^{1/2}}. \qquad \text{(A3.1.12)}$$

1. If σ is small, T_0 approaches

$$\frac{L}{n_0} - 1.$$

2. If σ is large, T_0 approaches

$$\frac{L^{1/2}}{\sigma}\left[\left(\frac{L}{n_0}\right)^{1/2} - \left(\frac{L}{n_0}\right)^{-1/2}\right].$$

In the first case when σ is small, few mistakes are made in selecting effective modifications. In this situation, as we might expect, the profit then depends very little on the actual size of the testing error σ and is proportional to $L/n_0 - 1$, which is the number of modifications tested.

In the second case, because σ is large, many mistakes are made in selecting modifications. Since the contribution from $(L/n_0)^{-1/2}$ will be small, profit will now increase only as the *square root* of the number of modifications tested. Furthermore, since now the profit is inversely proportional to σ, a sufficiently large value of σ will ensure that only a small profit will be made even if the best strategy is used.

When m is *positive*, it is to be noted that the optimal value $\xi_0 = -m\sigma^2/n$ is necessarily *negative*. This is at first sight surprising, because, for example, any significance test would certainly require the critical value to be positive. With this optimally chosen value of ξ it is still true, as in the case $m = 0$, that the best value of n is the smallest possible. However, this is not now true for all values of ξ.

When m is *negative*, it does not seem possible to draw any general conclusions, except that in this case the optimal value of ξ is *positive* and the optimal value of n may not now be the smallest possible.

Let us define a "significance test procedure" to mean one where acceptance or rejection of a modification is decided only after *replicated* trials

Table A3.1. Average profit for various values of m, σ^2, ξ and n

Testing Error Standard Deviation			$\sigma = \frac{1}{2}$					$\sigma = 1$					$\sigma = 2$				
Mean of Effect distribution	Number (n) of replications		20	10	4	2	1	20	10	4	2	1	20	10	4	2	1
	Number (p) of modifications tested		5	10	25	50	100	5	10	25	50	100	5	10	25	50	100
$m = 1$ Max. possible profit 107.3	Critical level ξ	4	0.0	0.0	0.2	0.4	1.3	0.0	0.0	0.3	1.1	4.6	0.1	0.3	1.1	4.6	16.1
		1	3.6	8.1	21.3	42.9	84.8	3.7	7.9	20.6	40.5	77.4	3.5	7.5	18.8	35.8	67.2
		0	4.3	9.7	25.8	52.3	104.3	4.4	9.6	25.3	50.3	97.0	4.2	9.3	23.5	44.7	82.6
		−1	4.1	9.3	24.8	50.7	102.5	4.2	9.3	24.8	50.7	101.5	4.1	9.3	24.6	48.7	92.5
		−4	4.0	9.0	24.0	49.0	99.0	4.0	9.0	24.0	49.0	99.0	4.0	9.0	24.0	49.1	99.2
	Optimal value of ξ		−0.0125	−0.025	−0.0625	−0.125	−0.25	−0.05	−0.10	−0.25	−0.50	−1.00	−0.20	−0.40	−1.00	−2.00	−4.00
$m = 0$ Max. possible profit 39.5	Critical level ξ	4	0.0	0.0	0.0	0.0	0.1	0.0	0.0	0.0	0.1	0.5	0.0	0.0	0.1	0.8	3.6
		1	1.0	2.2	5.8	11.8	23.7	0.8	1.8	5.7	11.4	21.8	0.8	1.8	5.3	9.6	16.0
		0	1.6	3.6	9.3	18.4	35.3	1.3	2.7	8.4	16.0	27.9	1.2	2.4	6.8	11.3	17.7
		−1	1.0	2.2	5.8	11.8	23.7	0.8	1.8	5.7	11.4	21.8	0.8	1.8	5.3	9.6	16.0
		−4	0.0	0.0	0.0	0.0	0.1	0.0	0.0	0.0	0.1	0.5	0.0	0.0	0.1	0.8	3.6
	Optimal value of ξ		0.0	0.0	0.0	0.0	0.0	0.0	0.0	0.0	0.0	0.0	0.0	0.0	0.0	0.0	0.0
$m = -1$ Max. possible profit 8.3	Critical level ξ	4	0.0	0.0	0.0	0.0	0.0	0.0	0.0	0.0	0.0	0.0	0.0	0.0	0.0	0.1	0.2
		1	0.1	0.3	0.8	1.7	3.5	0.2	0.3	0.8	1.7	2.5	0.1	0.3	0.6	−0.3	−6.5
		0	0.3	0.7	1.8	3.3	5.3	0.4	0.6	1.3	1.3	−2.0	0.2	0.3	−0.5	−4.3	−16.4
		−1	−0.4	−0.9	−2.7	−6.1	−14.2	−0.3	−1.1	−3.4	−8.5	−21.6	−0.5	−1.5	−5.2	−13.2	−31.8
		−4	−4.0	−9.0	−23.8	−48.6	−97.7	−4.0	−9.0	−23.7	−47.9	−95.4	−3.9	−8.7	−22.9	−44.4	−82.9
	Optimal value of ξ		0.0125	0.025	0.0625	0.125	0.25	0.05	0.10	0.25	0.50	1.00	0.20	0.40	1.00	2.00	4.00

have led to a *statistically significant* improvement at some level of probability (i.e., one where $n > 1$ and $\xi > 0$). Then, in the foregoing context and insofar as *empirical* feedback is concerned, we see that a significance testing procedure is justified only when we believe that most modifications will make matters worse.

For negative m, and to some extent when m is positive, the conclusions depend on the relative magnitude of terms in rather complicated expressions. To study the behavior of the solution under more general conditions calculations have been performed covering interesting ranges of the parameters. The results are presented in Table A3.1. We have supposed that the smallest possible value for n_0 is one time unit. (For a continuous process the length of this basic trial might be determined by the time it took for the process to settle down after modification. Alternatively, in a batch process, it could be the time taken to run a modified and an unmodified batch.) We suppose that L, the total time available, is 100 such time units. In this time we could run 100 modifications setting $n = 1$, 50 modifications setting $n = 2$, and so on. We consider the three values of m, $m = -1$, $m = 0$, and $m = 1$ corresponding roughly to the situations where only one-sixth of the modifications are improvements, where one-half of the modifications are improvements, and where five-sixths of the modifications are improvements. Three values of σ have been examined: $\sigma = \frac{1}{2}$, $\sigma = 1$, $\sigma = 2$. The case $\sigma = 2$ represents a particularly severe situation in the sense that the errors are very large compared with the sizes of the effects. Five values of the critical level ξ are considered: $\xi = -4, -1, 0, 1, 4$. Also computations are performed for five values of p, the number of modifications tested: $p = 5, 10, 25, 50$, and 100 corresponding to 20, 10, 4, 2, 1 replications of the basic comparison.

The quantity actually tabulated is

$$\frac{T}{\frac{1}{2}L} = (p - 1)\left[mF\left(\frac{\xi - m}{s}\right) + s^{-1}f\left(\frac{\xi - m}{s}\right)\right],$$

which, apart from the constant $\frac{1}{2}L$, is the expected profit.

On the left of the table is also shown the "maximum possible profit." This is the amount that could be obtained with $p = 100$ if σ were zero, and, consequently, there were no chance of misclassification.

In practice, of course, we will not know m or σ^2, at least not very precisely, and we can at best hope for a choice of ξ and n that will work fairly well over a wide range of values of m and σ^2. Inspection of the table shows that the smallest number of replications ($n = 1$) secures the highest rewards except in the least favorable situation considered ($\sigma = 2$, $m = -1$). In this last situation, whatever is done, very little can be gained but much can be lost. A small positive value for ξ (say $\xi = 1$, with $n = 1$) provides a useful hedge against the unfavorable possibilities without losing a very large proportion of what can be gained in favorable circumstances.

In general, when σ is very large and m is negative (so that the majority of modifications are unfavorable), it seems impossible to avoid the possibility of loss unless we know m or adopt a strategy which provides for learning about m as we proceed. Such a strategy will not be further discussed here. In particular, with m unknown, we will not be protected by a simple significance test, for substitution of $\xi = t\sigma/\sqrt{n}$ in (A3.1.6) can always be made to yield a negative result by suitable choice of a negative m.

Empirical and Scientific Feedback in Evolutionary Operation

There are two reasons for investigating the consequences of a particular set of assumptions:

1. Unassailable assumptions may lead to a change of opinion about consequences,
2. Unacceptable consequences may point up the inadequacy of assumptions.

The latter occurrence can be as instructive as the former and, indeed, is an important element in scientific progress.

It is obvious that the empirical feedback model considered here (even allowing for necessary simplification) is not appropriate for most actual applications of EVOP and still less for the process of general scientific inquiry. Its investigation is worthwhile, however, to make clear the necessity for a distinction between what may be called *empirical* feedback and *scientific* feedback. In empirical feedback an experimental result produces a simply related response in accordance with some definite rule. The empirical feedback rule investigated here is: If the observed effect of a modification to a production process is greater than ξ, include it as a permanent feature of the process, otherwise reject it. Scientific feedback, on the other hand, occurs as a result of an interaction between the data and the experience and knowledge of the experimenter. A particular pattern of changes in the composition could raise conjectures in the mind of a chemist resulting in experimental action which was quite unpredictable on any simple empirical rule.

The reason that we must carefully distinguish between these two kinds of feedback is because the strategy of experimentation which should be adopted depends on the type of feedback to be induced. We have just seen that, over a wide range of conditions, optimal empirical feedback is achieved when the maximum number of modifications are tested for the shortest possible time, ignoring completely the question of whether apparently favorable effects are due to chance.

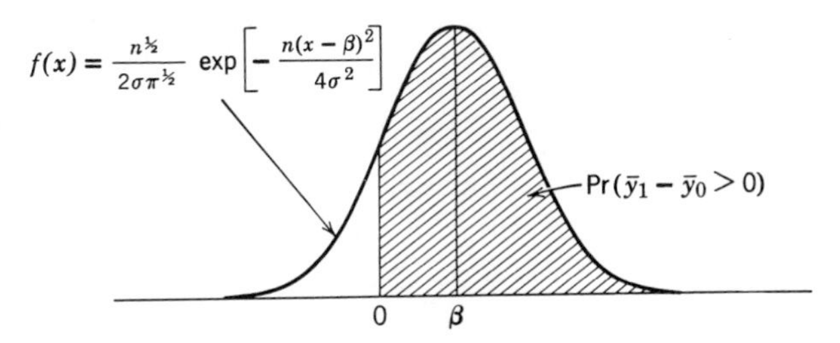

Figure A3.1. $\Pr(\bar{y}_1 - \bar{y}_0 > 0)$ when $\bar{y}_1 - \bar{y}_0 \sim N(\beta, 2\sigma^2/n)$.

By contrast, to provide effective scientific feedback it is necessary to have sufficient replication so that the effects about which the technologist is to reason are established with adequate precision. Anyone who has been exposed to experimental work knows how fruitless it is to build theories on the sand of inadequate data. A plausible theory can be produced to explain almost any set of data, even a set of random numbers.

Our own experience has been that most of the successes of EVOP have been achieved when it has been used as a means to learn more about the process. In this, scientific feedback has the most important although not the exclusive role. In particular, simplex EVOP can be expected to work well when empirical feedback is the only consideration, but not when, as is usually the case, scientific feedback is of major importance.

A3.2. CHOICE OF STRATEGY USING EMPIRICAL FEEDBACK TO ACHIEVE MAXIMUM INCREASE IN A RESPONSE *y* WHEN THE PROCESS FUNCTION IS A LINEAR FUNCTION OF A VARIABLE *w*

Suppose that the observed response y is a random variable with standard deviation σ and expected value

$$E(y) = \beta w, \tag{A3.2.1}$$

where w is an independent variable measured on some prescribed scale and where β is positive. Suppose we perform n trials at each of the values $w = 0$ and $w = 1$, and $\bar{y}_0$ and $\bar{y}_1$ are the corresponding average responses, so that approximately,

$$\bar{y}_1 - \bar{y}_0 \sim N\left(\beta, \frac{2\sigma^2}{n}\right). \tag{A3.2.2}$$

The chance that a positive effect will be observed and that we will move one step forward is (see Figure A3.1)

$$\Pr(\bar{y}_1 - \bar{y}_0 > 0) = \int_0^\infty \frac{n^{1/2}}{2\sigma\pi^{1/2}} \exp\left[-\frac{(x-\beta)^2 n}{4\sigma^2}\right] dx$$

$$= F\left(-\frac{\beta n^{1/2}}{\sigma 2^{1/2}}\right) = F, \qquad \text{(A3.2.3)}$$

where $F(-\beta n^{1/2}/\sigma 2^{1/2})$ is defined by (A3.1.1). The chance that a negative effect will be observed and we will move backwards one step is $1 - F$.

Suppose we score $+1$ for a forward step and -1 for a backward step; then the score measures progress in a forward (correct) direction. The expected score after $2N$ observations is given by

$$\frac{N}{n}[(+1)F + (-1)(1-F)] = \frac{N}{n}(2F-1)$$

$$= \frac{N}{n} 2 \int_0^{\beta n^{1/2}/\sigma 2^{1/2}} \frac{1}{(2\pi)^{1/2}} e^{-u^2/2}\, du. \qquad \text{(A3.2.4)}$$

If we consider the area under the $N(0, 1)$ distribution curve in the intervals $(0, \beta/\sigma 2^{1/2})$, $(\beta/\sigma 2^{1/2}, 2^{1/2}\beta/\sigma 2^{1/2})$, $(2^{1/2}\beta/\sigma 2^{1/2}, 3^{1/2}\beta/\sigma 2^{1/2})$, ..., we find that each area is clearly smaller than the one preceding it. Thus the average area in n successive intervals decreases as we increase n. Now the expected score (A3.2.4) is simply $2N$ (average area in n such intervals). Thus the expected score is maximized by making n as small as possible, that is, by selecting $n = 1$.

APPENDIX 4

How Many Cycles Are Necessary to Detect Effects of Reasonable Size?*

Results

In running an EVOP scheme, we inevitably introduce additional variation by changing the variables we want to study. Thus EVOP does not really get something for nothing. It obtains information from the process at the expense of slightly increasing its variability. Now it is the *proportional* increase in variability which determines the detectibility of the effects of the variables. It follows that we can obtain some idea of the number of cycles which might be needed for a given phase of an EVOP program by considering the proportional increase in the standard deviation of the response which would be acceptable. Normally, formal tests of significance at some fixed significance level are out of place in the routine running of EVOP. Nevertheless, for the purposes of making calculations, it becomes necessary to formalize our model, and in this investigation we have done this in terms of the familiar Neyman-Pearson theory of testing of hypotheses.

In Table A4.1 the quantity k is the proportional increase in standard deviation produced by changing the variables in accordance with the factorial design of the EVOP program. The table shows results for both the 2^2 and 2^3 factorial design. For the sake of simplicity, center points, which are often included in EVOP designs, are omitted. The quantities α and β are the usual errors of the first and second kind, α representing the chance of wrongly "detecting" the effects of the variables when no effects exist and β the risk of failing to detect effects which do exist. In our analysis, again for simplicity, we have supposed that only main effects occur. The entries in the body of the table indicate the number of cycles necessary to achieve

* See Section 7.1.

***Table A4.1.* Number of cycles required to detect (with probability $1 - \beta$, using an α-level test) main effects which increase the standard deviation of a process from σ to $k\sigma$ in two-factor and three-factor EVOP schemes without center points**

			k						
Design	α	β	1.2	1.3	1.4	1.5	1.6	1.7	1.8
	0.10	0.10	5.6	3.6	2.6	2.0	1.6	1.3	1.1
2^2	0.10	0.05	6.8	4.4	3.1	2.4	1.9	1.6	1.3
factorial	0.05	0.10	6.9	4.4	3.2	2.4	1.9	1.6	1.4
	0.05	0.05	8.2	5.2	3.8	2.9	2.3	1.9	1.6
	0.10	0.10	3.2	2.1	1.4	1.1	0.9	0.8	0.6
2^3	0.10	0.05	3.9	2.5	1.8	1.4	1.1	0.9	0.8
factorial	0.05	0.10	3.9	2.5	1.8	1.4	1.1	0.9	0.8
	0.05	0.05	4.6	3.0	2.1	1.6	1.3	1.1	0.9

the stated values of α and β. For information these numbers have been quoted (to one decimal place) in the manner in which they emerge from the appropriate calculations. Of course, the number of cycles must be an integer, so that suitable rounding should be performed where necessary.

The results indicate, for example, that when increases in standard deviation of 30% (i.e., $k = 1.3$) could be tolerated, then two or three cycles of three-variable EVOP programs should be enough to ensure small values of α and β. Of course, when more precise estimation is needed, a greater number of cycles will be required. The table indicates that, generally, a rather modest number of cycles should be adequate.

Technical Derivation of Table A4.1

Suppose p variables are being examined by means of a 2^p factorial design in one phase of an EVOP scheme; usually p will equal 2 or 3. Suppose, further, that the levels selected for these variables produce main effects $E_1, E_2, \ldots, E_p$ on a single response and that there are no interactions between the variables. Thus, for example, a change from the lower to the upper level of the ith variable produces an increase of E_i ($i = 1, 2, \ldots, p$) on the response, independent of the levels of other variables. Let e_i denote the estimate of E_i obtained from the average responses after n cycles in the usual manner and let σ_e^2 denote the variance of any of these effects, since all have equal variance.

Consider the hypothesis H_0 that all $E_i = 0$ against the alternative

H_1 that $E_i \neq 0$ for at least one i. If the random errors attached to the observations are $N(0, \sigma^2)$, then each e_i will also be normally distributed, in fact $e_i \sim N(E_i, \sigma_e^2)$, where $\sigma_e^2 = 4\sigma^2/n2^p$. Thus if H_0 is true,

$$u = \sum_{i=1}^{p} e_i^2 \sim \chi_p^2 \sigma_e^2, \tag{A4.1}$$

where χ_p^2 denotes a chi-square variable with p degrees of freedom. However if H_1 is true,

$$u = \sum_{i=1}^{p} e_i^2 \sim \chi_p^2(\Lambda)\sigma_e^2, \tag{A4.2}$$

where $\chi_p^2(\Lambda)$ denotes a noncentral χ^2 variable with p degrees of freedom and a noncentrality parameter

$$\Lambda = \sum_{i=1}^{p} \frac{E_i^2}{\sigma_e^2} \tag{A4.3}$$

[e.g., see Keeping (1962), page 397]. We shall now wish to select the value of n, the number of cycles, so that we can detect differences with a probability $1 - \beta$ while using an α-level test. This means that we shall require that, *if*

$$\frac{u}{\sigma_e^2} \geqq \chi_{p,1-\alpha}^2,$$

where the right-hand side denotes the $1 - \alpha$ point of the χ_p^2 distribution, leaving area α in the right tail, then also

$$\frac{u}{\sigma_e^2} \geqq \chi_{p,\beta}^2(\Lambda),$$

where the right-hand side is the β point of the $\chi_p^2(\Lambda)$ distribution, leaving area β in the left tail. This can be achieved if n is chosen so that

$$\chi_{p,1-\alpha}^2 = \chi_{p,\beta}^2(\Lambda). \tag{A4.4}$$

In practice a nonintegral value of n will be required to satisfy this equation, but the value can be suitably rounded. Larger values of n provide a higher certainty of detection at the same level α or the same certainty at a level smaller than α.

The noncentrality parameter. Suppose the variance of observations in the absence of effects is $V(y) = \sigma^2$. If a change in level of the ith variable produces an effect E_i in the response, the over-all variance increases by $E_i^2/4$. Since there are p effects in all, the over-all variance increases to

$$\sigma^2 + \frac{1}{4} \sum_{i=1}^{p} E_i^2, \tag{A4.5}$$

which by A(4.3) can be written as

$$\sigma^2\left(1 + \frac{1}{4}\Lambda\frac{\sigma_e^2}{\sigma^2}\right). \tag{A4.6}$$

If the variance is thus increased by a factor k^2 from σ^2 to $k^2\sigma^2$, then by setting (A4.6) equal to $k^2\sigma^2$ and using the fact that $\sigma_e^2 = 4\sigma^2/n2^p$, we obtain

$$\Lambda = n2^p(k^2 - 1). \tag{A4.7}$$

Thus effects which increase the standard deviation of a process from σ to $k\sigma$ produce a noncentrality parameter of a size given by (A4.7).

Exact solution of (A4.4).

1. For specified values of p, α, and β, look up in noncentral χ^2 tables the value $\Lambda = \Lambda_0$ which satisfies (A4.4).
2. Evaluate for any selected value of k,

$$n = \frac{\Lambda_0}{2^p(k^2 - 1)}.$$

This method is simple but possible only for $\alpha = 0.05$ and 0.01 and $\beta = 0.1(0.1)0.9$. The appropriate tables are given by Fix (1949). For other values of α and β an approximate method of calculation can be employed which appears to provide perfectly adequate accuracy.

Approximate Solution of (A4.4). A useful [Patnaik (1949)] approximation to the β percentage point, $\chi^2_{p,\beta}(\Lambda)$, of the noncentral χ^2 distribution with p degrees of freedom and noncentrality parameter Λ is given by $A\chi^2_{B,\beta}$, where

$$A = 1 + \frac{\Lambda}{p + \Lambda},$$
$$B = p + \frac{\Lambda^2}{p + 2\Lambda}, \tag{A4.8}$$

where Λ is given by (A4.7). This approximation converts (A4.4) into the (approximate) equality

$$\frac{p + n2^p(k^2 - 1)}{p + n2^{p+1}(k^2 - 1)}\chi^2_{p,1-\alpha} = \chi^2_{B,\beta}. \tag{A4.9}$$

The simplest way to solve (A4.9) for n when the values of α, β, p, and k are selected is to tabulate, for various (integer) values of n, both the left-hand side and the values of B which correspond. The values $\chi^2_{B,\beta}$ are then obtained by interpolating in the central χ^2 tables, since B will be nonintegral.

These calculations will then allow a nonintegral value of n to be interpolated which satisfies (A4.9). The results are as already given in Table A4.1.

For purposes of checking the approximation, we have evaluated the third and seventh lines of Table A4.1 by the exact method. The corresponding results are respectively:

7.2 4.6 3.3 2.5 2.0 1.7 1.4

and

4.0 2.6 1.8 1.4 1.1 0.9 0.8.

For the present purpose, the approximation appears to be adequate.

Tables

Table I. Normal distribution (single-sided).

Table II. Probability points of the normal distribution (single-sided).

Table III. Probability points of the normal distribution (double-sided).

Table IV. A table of factors w_k which convert the range of a normal sample of size k into an estimate of the standard deviation.

Table V. A table of values of $f_{k,n}$.

Table VI. Useful factors for EVOP calculations.

Table VII. A short table of random normal deviates.

Table I. Normal distribution (single-sided) [1]

Proportion (A) of whole area lying to right of ordinate through $x = \mu + z\sigma$

$$[z = (x - \mu)/\sigma]$$

Deviate (z)	Prefix	0.00	0.01	0.02	0.03	0.04	0.05	0.06	0.07	0.08	0.09	Prefix	Deviate (z)
0.0	0.5	000	960	920	880	840	801	761	721	681	641	0.4	0.0
0.1	0.4	602	562	522	483	443	404	364	325	286	247	0.4	0.1
0.2	0.4	207	168	129	090	052	013	974	936	897	859	0.3	0.2
0.3	0.3	821	783	745	707	669	632	594	557	520	483		0.3
0.4		446	409	372	336	300	264	228	192	156	121	0.3	0.4
0.5	0.3	085	050	015	981	946	912	877	843	810	776	0.2	0.5
0.6	0.2	743	709	676	643	611	578	546	514	483	451		0.6
0.7		420	389	358	327	296	266	236	206	177	148	0.2	0.7
0.8	0.2	119	090	061	033	005	977	949	922	894	867	0.1	0.8
0.9	0.1	841	814	788	762	736	711	685	660	635	611		0.9
1.0		587	562	539	515	492	469	446	423	401	379		1.0
1.1		357	335	314	292	271	251	230	210	190	170	0.1	1.1
1.2	0.1	151	131	112	093	075	056	038	020	003	985	0.0	1.2
1.3	0.0	968	951	934	918	901	885	869	853	838	823		1.3
1.4		808	793	778	764	749	735	721	708	694	681		1.4
1.5		668	655	643	630	618	606	594	582	571	559		1.5
1.6		548	537	526	516	505	495	485	475	465	455		1.6
1.7		446	436	427	418	409	401	392	384	375	367		1.7
1.8		359	351	344	336	329	322	314	307	301	294		1.8
1.9		287	281	274	268	262	256	250	244	239	233		1.9
2.0		228	222	217	212	207	202	197	192	188	183		2.0
2.1		179	174	170	166	162	158	154	150	146	143		2.1
2.2		139	136	132	129	125	122	119	116	113	110	0.0	2.2
2.3	0.0	107	104	102	990	964	939	914	889	866	842	0.00	2.3
2.4	0.00	820	798	776	755	734	714	695	676	657	639		2.4
2.5		621	604	587	570	554	539	523	508	494	480		2.5
2.6		466	453	440	427	415	402	391	379	368	357		2.6
2.7		347	336	326	317	307	298	289	280	272	264		2.7
2.8		256	248	240	233	226	219	212	205	199	193		2.8
2.9	0.00	187	181	175	169	164	159	154	149	144	139	0.00	2.9

[1] Adapted from Davies (1956), condensed and adapted from Pearson and Hartley (1954).

Table I (continued). Extension for higher values of the deviate [1]

Deviate (z)	Proportion of Whole Area (A)	Deviate (z)	Proportion of Whole Area (A)	Deviate (z)	Proportion of Whole Area (A)	Deviate (z)	Proportion of Whole Area (A)
3.0	0.00135	3.5	0.000233	4.0	0.0^4317	4.5	0.0^5340
3.1	0.000968	3.6	0.000159	4.1	0.0^4207	4.6	0.0^5211
3.2	0.000687	3.7	0.000108	4.2	0.0^4133	4.7	0.0^5130
3.3	0.000483	3.8	0.0^4723	4.3	0.0^5854	4.8	0.0^6793
3.4	0.000337	3.9	0.0^4481	4.4	0.0^5541	4.9	0.0^6479
						5.0	0.0^6287

[1] Adapted from Davies (1956), condensed and adapted from Pearson and Hartley (1954).

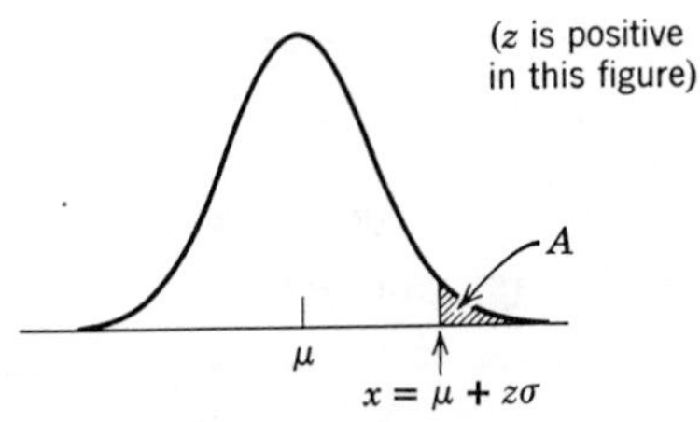

The illustration shows a normal curve. The scales are such that the total area under the curve is unity. The shaded portion is the area A given in Table I. The entries refer to positive values of the argument z. For negative values of z write down the complements of the entries.

Examples. *Let* $z = +1.96$: The prefix = 0.0 and the entry = 250, together 0.0250 = area to right. Area to left = 1 − 0.0250 = 0.9750.

Let $z = -3.00$: The tabulated value = 0.00135. Since z is negative, this represents the area to the *left*. Area to right = 1 − 0.00135 = 0.99865.

Let $z = +4.50$: Tabulated value = 0.00000340. Area to left = 0.99999660.

To find the value of z corresponding to a given A, we can use Table I in reverse, thus:

Let area to right (i.e., A) = 0.10: The two adjacent tabulated values are $A = 0.1003$ for $z = 1.28$, and $A = 0.0985$ for $z = 1.29$. We interpolate linearly to obtain the required value of z. Thus $z = 1.28 + (3)(0.01)/18 = 1.2817$. For a slightly more accurate value and for certain other pairs of values of A and z, see Table II.

Table II. Probability points of the normal distribution (single-sided) [1]

$$[z = (x - \mu)/\sigma]$$

A	z
0.001	3.0902
0.005	2.5758
0.01	2.3263
0.02	2.0537
0.025	1.9600
0.05	1.6449
0.1	1.2816
0.2	0.8416
0.3	0.5244
0.4	0.2533
0.5	0.0000

[1] Adapted from Davies (1956), condensed and adapted from Fisher and Yates (1948).

Table III. Probability points of the normal distribution (double-sided) [1]

$$[z = (x - \mu)/\sigma]$$

P	z
0.001	3.2905
0.005	2.8070
0.01	2.5758
0.02	2.3263
0.025	2.2414
0.05	1.9600
0.1	1.6449
0.2	1.2816
0.3	1.0364
0.4	0.8416
0.5	0.6745

[1] Adapted from Davies (1956), condensed and adapted from Fisher and Yates (1948).

Table II

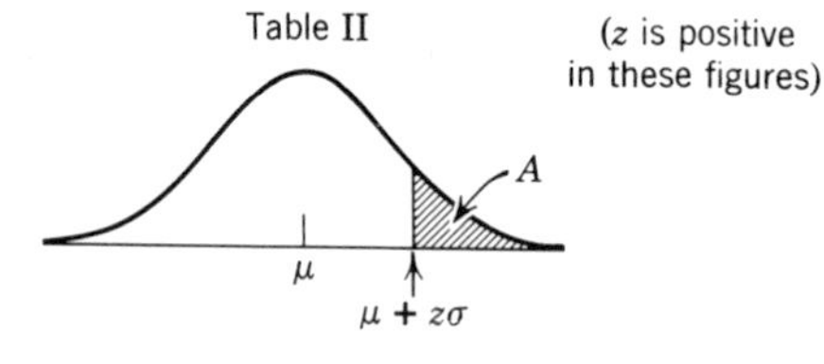

(z is positive in these figures)

Table III

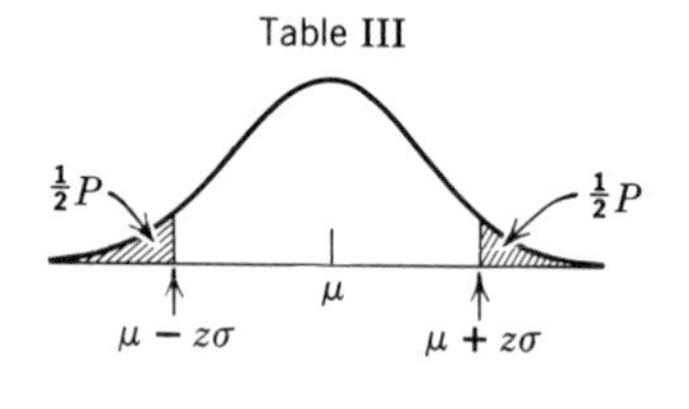

The illustrations show two normal curves. The scales are such that the total area under each curve is unity. The shaded areas under the left and right curves are the arguments A and P in Table II and III respectively.

Examples. The deviate of a normal curve which cuts off a single tail equivalent to $A = 0.05$ is given by $z = 1.6449$. Alternatively, the chance that an item taken at random from a normal population specified by mean μ and standard deviation σ will not exceed in the algebraic sense $\mu + 1.6449\sigma$ is 0.95, or 95%.

The deviates of a normal curve which jointly cut off a double tail equivalent to $P = 0.05$ are given by $z = \pm 1.9600$. Alternatively, the chance that an item taken at random from a normal population specified by mean μ and standard deviation σ will not deviate from μ *in either direction* by more than 1.9600σ is 0.95, or 95%.

Table IV. ***A table of factors w_k which convert the range of a normal sample of size k into an estimate of the standard deviation****

k	w_k	k	w_k
2	0.8862	11	0.3152
3	0.5908	12	0.3069
4	0.4857	13	0.2998
5	0.4299	14	0.2935
6	0.3946	15	0.2880
7	0.3698	16	0.2831
8	0.3512	17	0.2787
9	0.3367	18	0.2747
10	0.3249	19	0.2711
		20	0.2677

* Taken from Pearson, E. S. and H. O. Hartley, *Biometrika Tables*, Vol. I, Cambridge University Press, England, 1966, p. 177.

Table V. ***A table of values of $f_{k,n}$***

No. of Cycles,	Number of Runs in the Block, k								
n	2	3	4	5	6	7	8	9	10
2	0.63	0.42	0.34	0.30	0.28	0.26	0.25	0.24	0.23
3	0.72	0.48	0.40	0.35	0.32	0.30	0.29	0.27	0.26
4	0.77	0.51	0.42	0.37	0.34	0.32	0.30	0.29	0.28
5	0.79	0.53	0.43	0.38	0.35	0.33	0.31	0.30	0.29
6	0.81	0.54	0.44	0.39	0.36	0.34	0.32	0.31	0.30
7	0.82	0.55	0.45	0.40	0.37	0.34	0.33	0.31	0.30
8	0.83	0.55	0.45	0.40	0.37	0.35	0.33	0.31	0.30
9	0.84	0.56	0.46	0.40	0.37	0.35	0.33	0.32	0.31
10	0.84	0.56	0.46	0.41	0.37	0.35	0.33	0.32	0.31
15	0.86	0.57	0.47	0.42	0.38	0 36	0.34	0.33	0.31
20	0.86	0.58	0.47	0.42	0.38	0.36	0.34	0.33	0.32

Table VI. Useful factors for EVOP calculations

n	$\frac{1}{n}$	$\frac{1}{n-1}$	$\frac{1}{2(n-1)}$	$\frac{2}{\sqrt{n}}$	$\frac{1.79}{\sqrt{n}}$	$\frac{1.73}{\sqrt{n}}$	$\frac{1.41}{\sqrt{n}}$	$\frac{1.26}{\sqrt{n}}$
1	1.00	—	—	2.00	1.79	1.73	1.41	1.26
2	0.50	1.00	0.50	1.41	1.26	1.22	1.00	0.89
3	0.33	0.50	0.25	1.15	1.03	1.00	0.82	0.73
4	0.25	0.33	0.17	1.00	0.89	0.87	0.71	0.63
5	0.20	0.25	0.12	0.89	0.80	0.77	0.63	0.57
6	0.17	0.20	0.10	0.82	0.73	0.71	0.58	0.52
7	0.14	0.17	0.08	0.76	0.68	0.65	0.53	0.48
8	0.12	0.14	0.07	0.71	0.63	0.61	0.50	0.45
9	0.11	0.12	0.06	0.67	0.60	0.58	0.47	0.42
10	0.10	0.11	0.06	0.63	0.57	0.55	0.45	0.40

Table VII. A short table of random normal deviates

−0.26	−0.11	−0.42	−0.41	0.46	1.09	−0.58	−0.43	1.12	−0.06
0.58	−0.46	0.76	−1.89	0.03	0.05	0.18	−0.14	−0.74	−1.04
0.91	−0.74	0.13	−1.01	−0.49	−0.38	0.55	0.64	0.03	0.38
0.17	1.72	0.72	−1.19	1.83	1.22	−0.19	0.83	−0.23	−0.32
0.11	1.36	−0.22	0.17	1.15	−0.76	−0.16	−1.14	1.33	−1.18
−0.63	0.23	−0.43	1.36	2.40	1.66	0.16	0.20	−0.97	0.94
1.67	1.13	−1.44	0.27	0.99	0.56	−2.08	−0.07	−0.04	0.58
−0.90	0.47	−0.18	−0.51	1.70	−0.58	1.16	−0.13	0.42	−1.05
−1.29	0.62	−0.65	0.73	−0.23	−0.83	−0.96	1.07	2.37	1.44
−0.59	−2.43	−1.00	−1.10	−0.25	−0.30	0.86	−0.05	−0.70	1.28
−0.37	0.74	1.63	−0.52	−0.01	1.01	−0.61	2.28	0.89	−0.29
0.87	−0.60	−1.70	2.38	0.01	−1.66	0.03	0.53	0.98	−1.05
−0.01	−1.01	0.39	0.75	0.28	−0.40	2.31	−0.49	0.35	0.11
−1.39	−1.65	0.47	−2.26	0.33	1.52	0.55	2.32	0.62	0.24
−0.56	−0.60	−0.12	0.59	−0.29	−1.63	0.68	−0.17	−0.56	−0.67
1.26	−0.75	1.90	0.43	−1.75	−1.30	1.64	−0.80	−0.33	−1.98
−2.55	−0.03	−1.11	−0.38	0.23	0.41	−0.48	1.06	0.61	−0.20
1.55	0.45	0.24	0.58	1.56	0.25	−1.91	0.90	−0.58	0.11
−0.30	0.68	2.17	−0.90	0.88	1.39	−0.34	−0.05	0.94	−0.14
−0.56	−2.07	−1.25	0.18	1.68	1.32	3.39	1.21	−1.72	−0.29
−0.24	−0.47	−0.53	0.83	0.55	1.85	2.23	−1.16	0.12	−0.16
−0.16	−2.59	0.30	−1.22	−0.79	−0.12	−0.25	−0.01	−0.25	0.09
−1.34	0.60	−0.27	−0.20	−0.17	1.83	−1.80	−0.04	−1.84	0.16
0.14	2.09	−0.30	0.26	−0.15	−0.95	−1.32	0.57	−0.13	0.32
−0.16	−2.51	−1.64	−0.88	0.92	1.54	−0.85	1.43	−1.57	−2.22
−0.90	−0.26	0.97	1.38	−0.02	1.41	−0.00	−0.79	0.14	−0.45
0.53	0.44	−2.50	−1.11	0.93	−0.18	−0.23	−0.37	0.38	1.39
0.30	−1.43	0.32	0.95	1.35	1.24	1.42	−0.20	0.71	−0.32
−2.03	−0.52	0.07	−0.84	0.63	0.68	0.53	−0.65	0.54	0.01
0.44	−0.11	0.58	−0.25	1.92	−1.41	−0.56	0.19	−0.86	−0.31
−0.87	0.18	1.17	−1.68	0.97	−0.85	−0.22	0.75	1.20	0.01
−1.85	−0.33	0.98	−2.24	−0.76	1.42	1.16	−1.00	3.20	1.18
1.18	0.27	−2.44	2.29	0.01	−1.34	0.21	−0.48	−0.69	0.50
1.15	−1.22	−1.78	−0.01	−0.36	0.11	−1.03	2.08	0.85	0.58
−2.01	−0.07	−1.15	−0.97	−0.11	0.24	−0.32	−0.51	−0.51	−0.01
1.11	−0.71	−0.52	−1.27	−0.21	−1.60	0.11	0.52	−1.63	−0.78
−0.50	−0.07	0.03	1.24	−0.40	1.22	−0.91	0.33	−1.66	1.52
0.51	0.72	1.42	2.08	−0.48	−0.04	−0.27	−1.29	−0.24	−0.61
−0.17	0.36	0.32	−0.68	−0.68	−2.38	−0.56	−2.77	−1.71	−1.04
−0.73	−0.71	−0.51	0.70	1.12	0.17	−1.54	−0.94	0.31	−0.17

Table VII (continued)

−0.19	0.56	2.45	−0.36	−0.44	0.58	0.79	1.67	0.77	−0.16
−0.37	−0.13	−0.24	−1.17	−1.84	−0.51	0.59	−1.91	0.68	0.41
−0.91	1.14	0.54	−1.92	−1.29	−1.86	−1.21	2.05	0.92	0.40
1.57	−0.91	0.66	0.21	0.04	−0.83	−1.43	−0.38	−0.22	−0.75
−1.20	0.16	−0.83	0.64	0.01	−0.84	1.47	−0.29	−1.23	1.49
−0.11	1.82	0.99	0.04	0.06	0.75	−0.89	1.46	−1.76	0.44
0.87	−0.29	−0.70	1.02	1.49	−1.00	1.32	1.25	−0.98	−0.39
0.91	1.20	1.19	−0.84	−0.07	0.48	−1.66	0.50	−1.66	−0.95
0.50	1.26	−0.10	3.74	0.62	0.69	0.01	−0.07	0.85	−0.79
−0.04	−1.10	−1.31	−0.75	0.53	0.67	0.29	−2.52	−2.03	1.56
−1.00	0.42	−0.70	0.15	0.37	0.41	0.29	0.48	1.00	1.47
0.25	−0.90	1.28	−0.66	0.18	−1.40	1.81	0.72	0.84	−0.24
−0.86	2.08	−0.25	0.74	−0.19	−1.10	0.70	0.02	0.20	0.09
1.71	−1.52	0.26	0.27	−0.39	−0.56	−0.69	−1.32	−0.81	0.95
0.50	1.56	−0.73	0.02	−1.61	−1.04	−0.60	0.67	−0.81	−1.89
−0.30	−1.74	−0.29	−1.35	−0.38	−0.82	1.29	0.42	−0.26	0.81
−1.18	−1.15	−0.06	−0.59	1.59	−0.25	0.36	0.10	−0.41	0.97
0.89	0.36	0.48	1.07	0.03	−0.44	−0.12	−0.34	0.83	0.18
0.00	−1.25	−1.71	−0.99	0.57	−0.06	−0.33	−0.58	−0.03	−0.20
−0.18	−0.17	−0.77	−0.80	0.17	0.80	0.12	−0.51	−0.64	−0.30
0.98	−0.23	−0.19	0.19	0.99	−2.01	−1.05	0.60	1.99	0.31
0.33	0.79	0.01	−0.24	2.08	−2.61	−0.15	−1.07	−1.67	0.10
−0.80	−0.62	0.95	0.90	0.47	0.75	−0.60	0.34	−0.12	−0.09
1.15	−0.82	−0.99	−1.17	−0.11	−2.20	−1.72	−0.12	−0.68	−0.24
−0.19	0.68	2.51	0.74	−0.47	−0.19	0.09	−1.51	−0.23	0.34
−1.09	0.39	0.87	0.63	1.29	−1.57	0.50	−1.05	−0.67	1.19
0.89	0.44	−0.61	0.03	−1.67	−1.49	0.02	−0.62	−0.86	−0.02
0.07	−0.04	0.21	0.71	−0.81	0.37	−0.58	−0.36	−0.49	0.33
0.00	1.60	3.10	−1.91	−0.35	−1.37	−0.08	1.06	−0.10	−1.03
−0.74	1.03	−0.88	−0.39	−1.71	−0.35	−0.21	−1.29	2.24	2.14
0.15	−0.83	0.22	1.03	−1.25	−2.39	1.58	−0.20	−0.08	−2.04
−0.62	−3.98	0.07	1.18	0.40	−1.75	0.40	1.31	−2.16	−1.20
0.33	−0.41	0.57	1.20	0.68	−0.89	0.48	−0.93	0.41	−1.04
−1.97	−1.13	−1.34	−0.32	−0.55	0.77	−0.62	0.24	−0.21	−2.00
2.52	−0.83	−0.12	1.00	1.04	0.34	−0.37	0.92	−2.39	0.79
0.65	−1.49	−1.10	0.67	1.34	1.83	−1.10	0.04	−1.58	−0.45
−0.35	0.51	−1.11	0.99	−0.67	0.65	−0.77	−0.02	−0.77	−0.29
0.42	0.78	−0.45	0.60	0.76	−0.33	3.82	0.64	−0.58	1.22
1.22	−0.73	−0.55	1.79	−2.23	−0.12	0.31	−0.15	1.24	−1.26
1.17	−0.75	0.31	1.20	−0.08	0.32	0.91	0.74	0.19	0.65

Table VII (continued)

−0.62	−0.92	−0.26	−0.49	−0.45	0.60	−0.68	−0.99	0.20	1.30
1.24	0.23	−0.86	−0.82	−0.01	1.13	0.56	0.76	1.26	0.00
−0.67	1.14	0.13	−0.64	1.50	0.21	0.21	−0.99	−0.14	0.02
−0.04	0.36	−0.50	−0.81	0.50	−0.74	0.70	0.51	1.53	0.54
−0.22	−0.35	0.14	1.32	0.41	−0.33	1.94	−0.83	0.64	−0.39
0.98	0.10	−0.23	2.68	0.07	−0.27	1.57	1.20	−0.22	1.26
0.24	0.61	−2.08	−1.21	1.20	−0.13	−0.71	0.47	−0.49	0.14
−0.52	0.99	0.22	0.94	0.77	0.06	−1.68	1.78	0.38	−0.10
1.18	−0.42	−0.32	−0.70	2.23	−0.01	−1.67	1.32	0.03	−1.12
0.59	−0.57	0.52	−0.64	0.53	−0.65	−2.47	2.03	−1.78	−1.55
0.74	−0.23	−1.88	−0.72	0.65	−1.02	0.37	0.74	−0.74	−0.35
1.17	−0.57	−0.53	0.15	0.90	−0.02	−0.64	−0.95	0.02	0.22
−2.25	0.17	0.58	−0.70	−1.32	0.06	0.79	0.07	1.16	0.68
0.49	0.12	1.37	−0.48	−0.06	−0.67	1.04	−0.30	−0.18	0.01
0.29	−0.34	−1.48	0.10	2.52	1.53	0.75	0.74	0.36	1.16
−0.25	−1.03	−0.36	0.03	−0.09	0.39	−0.59	−0.30	1.66	−0.52
−0.01	−0.40	1.27	−1.31	−0.27	0.35	−0.41	0.13	1.03	1.50
0.45	0.93	0.60	0.45	−0.54	−1.42	−2.04	−0.64	−0.48	0.39
2.47	0.62	−1.01	0.74	−0.30	0.17	−0.06	−0.49	−1.24	0.75
−1.15	1.29	0.03	−1.06	−0.73	−0.72	0.80	0.24	−0.42	1.19
2.04	1.31	0.75	−0.68	0.94	−0.02	0.02	0.47	−0.78	0.41
−0.37	−1.24	0.39	1.39	−1.28	0.58	−0.58	0.65	−0.63	0.40
1.41	1.90	−0.63	0.17	−0.61	−0.28	0.13	0.94	0.91	−0.73
0.41	1.81	−1.73	−0.15	0.99	0.17	−0.11	0.56	−0.52	−0.99
0.45	0.36	−0.86	1.26	−0.17	1.49	0.31	−3.07	1.07	−0.16
−0.28	−1.46	−0.51	0.29	−1.62	−0.49	0.14	1.82	−0.40	−0.17
−1.29	0.26	−1.68	−2.47	−1.72	0.44	−0.18	0.63	0.68	−0.54
−0.20	0.11	1.24	−0.68	0.50	1.68	0.46	−0.73	−0.06	0.07
−0.20	−0.75	0.44	1.08	−0.58	−1.31	−1.01	−0.85	1.91	1.46
−0.45	0.46	−1.39	0.81	0.11	−0.89	−0.28	−0.22	−0.21	0.40
0.03	2.38	0.17	−0.70	−0.44	0.46	−1.06	0.82	0.18	−0.99
0.22	1.95	−0.19	1.73	0.47	−1.10	0.45	1.39	−0.27	−1.04
−1.63	−1.81	−1.76	−1.06	−1.33	0.36	−1.03	0.64	0.09	−0.76
−0.16	−0.89	0.95	−0.55	1.54	−1.77	−0.86	−0.40	−0.61	2.45
0.85	0.17	−0.48	0.01	0.33	0.76	0.39	−0.66	−0.72	−1.13
−1.15	2.22	−0.67	0.39	1.64	0.67	0.47	0.24	1.07	0.65
−1.46	−0.69	−0.25	1.67	−0.37	0.26	−0.22	1.06	0.66	−1.65
1.54	−0.51	0.75	0.79	0.09	0.32	−0.96	−0.40	0.66	−0.61
−1.61	0.29	−0.88	−0.76	−0.87	1.88	−0.59	−2.87	0.66	−1.21
0.32	−1.05	−0.02	1.37	−0.43	1.90	0.72	2.22	0.76	−0.31

References and Bibliography

Anonymous (1959), Turning minor changes into major savings. *Chemical Week*, **85,** 75–77 (October 24).

Anonymous (1960), In big computer test, plant raises NH_3 output. *Chemical Engineering*, **67,** 110–114 (November 14).

Anonymous (1961*a*), EVOP in plant-scale experiments. *Industrial and Engineering Chemistry*, **53,** No. 12, 40A–46A.

Anonymous (1961*b*), Evolutionary operation: A tool to optimize a process without disturbing it. *Petroleum Week*, 48–49 (April 7).

Anonymous (1961*c*), Evolutionary operation. *Paper Mill News*, 10–11 (October 2).

Anonymous (1963), Maumee moves closer to optimum process to make saccharin. *Chemical and Engineering News*, 76–78 (December 9).

Baasel, W. D. (1965), Exploring response surfaces to establish optimum conditions. *Chemical Engineering*, **72,** 147–152 (October 25).

Barnett, E. H. (1960), Introduction to EVOP. *Industrial and Engineering Chemistry*, **52,** No. 6, 500–503

Barnett, E. H. (1961), Education for EVOP. Presented at 139th ACS Meeting, St. Louis, Missouri, March 21.

Beyer, W. H. (1968), *Handbook of Tables for Probability and Statistics* (2nd ed.). The Chemical Rubber Co., 18901 Cranwood Parkway, Cleveland, Ohio, 44128.

Bingham, R. S., Jr. (1962), Chalkdust and chemicals—S. Q. C. training for chemical industry. *Industrial Quality Control*, **18,** No. 10, 15–18.

Bingham, R. S., Jr. (1963), Try EVOP for systematic process improvement. *Industrial Quality Control*, **20,** No. 3, 17–23.

Box, G. E. P. (1954*a*), Some theorems on quadratic forms applied in the study of analysis of variance problems, II. Effects of inequality of variance and of correlation between errors in the two-way classification. *Annals of Mathematical Statistics*, **25,** 484–498.

Box, G. E. P. (1954*b*), The exploration and exploitation of response surfaces: Some general considerations and examples. *Biometrics*, **10,** 16–60.

Box, G. E. P. (1957), Evolutionary operation: a method for increasing industrial productivity. *Applied Statistics*, **6,** 81–101.

Box, G. E. P. (1960), Some general considerations in process optimization. *Journal of Basic Engineering*, March, 113–119.

Box, G. E. P. (1966*a*), A simple system of evolutionary operation subject to empirical feedback. *Technometrics*, **8,** 19–26.

Box, G. E. P. (1966*b*), Use and abuse of regression. *Technometrics*, **8**, 625–629.

Box, G. E. P., and J. Chanmugam (1962), Adaptive optimization of continuous processes. *Industrial and Engineering Chemistry Fundamentals*, **1**, 2–16.

Box, G. E. P., and G. A. Coutie (1956), Application of digital computers in the exploration of functional relationships. *Proceedings of the Institution of Electrical Engineers*, **103**, Part B, Supplement No. 1, 100–107; discussion 108–111.

Box, G. E. P., and N. R. Draper (1968), Isn't my process too variable for EVOP? *Technometrics*, **10**, 439–444.

Box, G. E. P., and J. S. Hunter (1959), Condensed calculations for evolutionary operation programs. *Techometrics*, **1**, 77–95.

Box, G. E. P., and J. S. Hunter (1961), The 2^{k-p} fractional factorial designs, Part I. *Technometrics*, **3**, 311–351; Part II, 449–458.

Box, G. E. P., and G. M. Jenkins (1962), Some statistical aspects of adaptive optimization and control. *Journal of the Royal Statistical Society, Series B*, **24**, 297–343.

Box, G. E. P., and G. M. Jenkins (1969), *Time Series, Forecasting, and Control*. Holden-Day, San Francisco.

Box, G. E. P., and K. B. Wilson (1951), On the experimental attainment of optimum conditions. *Journal of the Royal Statistical Society, Series B*, **13**, 1–45.

Brooks, S. H. (1958), A discussion of random methods for seeking maxima. *Operations Research*, **6**, 244–251.

Brooks, S. H. (1959), A comparison of maximum-seeking methods. *Operations Research*, **7**, 430–457.

Carpenter, B. H., and H. C. Sweeney (1965), Process improvement with "simplex" self-directing evolutionary operation. *Chemical Engineering*, **72**, 117–126 (July 5).

Cestoni, D. J., R. E. Ringelman, and L. R. Olson (1960), Process engineering a petrochemical plant. *Chemical Engineering Progress*, **56**, No. 5, 73–77.

Charnes, A., W. W. Cooper, and A. Henderson (1953), *An Introduction to Linear Programming*. Wiley, New York.

Chatto, K. A., and R. W. Kennard (1961*a*), EVOP in plant scale experiments—the simplified concepts. *Industrial and Engineering Chemistry*, **53**, No. 12, 47A, 48A, 52A.

Chatto, K. A., and R. W. Kennard (1961*b*), Administrative and statistical prerequisites for successful EVOP. Presented at 139th ACS Meeting, St. Louis, Missouri, March 22.

Chemstrand Corporation (1959), *Annual Report*. 13.

Cochran, W. G., and G. M. Cox (1957), *Experimental Designs*. Wiley, New York.

Coutie, G. A. (1959*a*), Use statistics for optimization. *Chemical Engineering*, **66**, 190–191 (November 16).

Coutie, G. A. (1959*b*), Statistical considerations in process optimization. *Proceedings of the Joint Symposium on Instrumentation and Computation in Process Development and Plant Design*, London, 1.

David, H. A. (1951), Further applications of range to the analysis of variance. *Biometrika*, **38**, 393–409.

Davies, O. L., editor (1956), *The Design and Analysis of Industrial Experiments* (2nd ed.). Oliver and Boyd, Edinburgh.

Davies, O. L., editor (1957), *Statistical Methods in Research and Production* (3rd ed.). Oliver and Boyd, Edinburgh.

Davison, I. D. (1961), EVOP—a new strategy in experimental design. *The B.H.P.* Tech*nical Bulletin*, Australia, **5**, 6–11.

DeBusk, R. E. (1962), Experience in EVOP at Tennessee Eastman Co. *Industrial Quality Control*, **19**, No. 4, 15–21.

Delver, A. (1961), Gradual optimalization of production processes, II. *SIGMA*, **7**, No. 3, 45–48.

(Abstracted by M. J. Maurice (1962) in *Quality Control and Applied Statistics*, **7,** 269.)

Draper, C. S., and Y. T. Li (1951), *Principles of Optimalizing Controls and an Application to the Internal Combustion Engine*. A.S.M.E., New York.

Draper, N. R., and H. Smith (1966), *Applied Regression Analysis*. Wiley, New York.

Enrick, N. L. (1963), *Cases in Management Statistics*. Holt, Rinehart, and Winston, New York.

Enrick, N. L. (1966), *Quality Control*, 5th Edition. Industrial Press, New York.

Filice, B. (1963), EVOP study of spinach blanching. Seminar on Evolutionary Operation, Sponsored by the National Canners Assoc. and ASQC, San Francisco, June 27.

Fisher, R. A., and F. Yates (1948), *Statistical Tables for Biological, Agricultural and Medical Research*. Oliver and Boyd, Edinburgh. (Sixth edition, 1963).

Fix, Evelyn (1949), Tables of non-central χ^2. *University of California Publications in Statistics*, **1,** No. 2, 15–19.

Grohskopf, H. (1960), Statistics in the chemical process industries—present and future. *Industrial and Engineering Chemistry*, **52,** No. 6, 497–499.

Grundy, P. M., M. J. R. Healy, and D. H. Rees (1956), Economic choice of the amount of experimentation. *Journal of the Royal Statistical Society, Series B*, **18,** 32–55.

Hadley, G. (1962), *Linear Programming*. Addison-Wesley, Reading, Massachusetts.

Hadley, G. (1964), *Nonlinear and Dynamic Programming*. Addison-Wesley, Reading, Massachusetts.

Hehner, H. O. (1963*a*), EVOP spoken here. Paper read at the 18th Midwest Quality Control Conference, Tulsa, Oklahoma, October 11.

Hehner, H. O. (1963*b*), EVOP deglamorized. Paper read to the Seventh Technical Conference of the Chemical Division of ASQC, St. Paul, Minnesota, November 7.

Hehner, H. O. (1963*c*), Personal communication, December 10, to W. G. Hunter.

Hehner, H. O., and E. H. Barnett (1963), Evolutionary operation—an annotated bibliography. Presented at the Seventh Technical Conference of the Chemical Division of ASQC, St. Paul, Minnesota, November 7.

Hicks, C. R. (1964), *Fundamental Concepts in the Design of Experiments*. Holt, Rinehart, and Winston, New York, 245–254.

Hunter, J. S. (1958), Some statistical principles underlying EVOP. Second Stevens Symposium Papers, 63–76, (January 25).

Hunter, J. S. (1960), Optimize your process with EVOP. *Chemical Engineering*, **67,** 193–202 (September 19).

Hunter, W. G., and K. Jayachandran (1966), Evolutionary operation and its basic principles. *Productivity* (India), **7,** No. 3, 377–392 (Autumn).

Hunter, W. G., and J. R. Kittrell (1966), Evolutionary operation: a review. *Technometrics*, **8,** 389–397.

Jenkins, G. M., and D. G. Watts (1968), *Spectral Analysis and Its Applications*, Holden-Day, San Francisco.

Keeping, E. S. (1962), *Statistical Inference*. Van Nostrand, New York.

Kenworthy, I. C. (1967), Some examples of simplex evolutionary operation in the paper industry. *Applied Statistics*, **16,** 211–224.

Kitagawa, T. (1962), A mathematical formulation of the EVOP programs. *Bulletin of the Institute of International Statistics*, Paris, **39,** No. 2, 293–309.

Klingel, A. R., and R. G. McIntyre (1962), An experimental strategy for investigating commercial processes. *Applied Statistics*, **11,** No. 2, 79–92.

Koehler, T. L. (1958*a*), Evolutionary operation, some actual examples. Second Stevens Symposium Papers, 5-8 (January 25).

Koehler, T. L. (1958*b*), Evolutionary operation: a program for optimizing plant operation.

Transactions All-Day Rutgers Quality Control Conference, American Society for Quality Control, 25–34 (September). [Abstracted by W. J. Chelgren (1959) in *Quality Control and Applied Statistics*, **4,** 453.]

Koehler, T. L. (1959*a*), EVOP. *Chemical Engineering Progress*, **55,** No. 10. 76–79.

Koehler, T. L. (1959*b*), Evolutionary operation: its methods and application. *Tappi*, **42,** No. 3, 261–264.

Koehler, T. L. (1960), How statistics apply to chemical processes. *Chemical Engineering*, **67,** 142–152 (December 12).

Koehler, T. L. (1965), EVOP. Fifteenth Anniversary Issue of the Japanese I. Q. C., *Hinhitsu Kanri* (in Japanese).

Koehler, T. L., EVOP: estimation of error variance. An unpublished paper.

Koleff, A. M. (1963), Application of EVOP to corn grinding operation at A. E. Staley Manufacturing Co. Personal communication, January 4, to W. G. Hunter.

Konz, S. A. (1965), Selecting speed and feed under factory conditions. *Tool and Manufacturing Engineer*, **55,** July, 31–33.

Kotnour, K. D., G. E. P. Box, and R. J. Altpeter (1966), A discrete predictor-controller applied to sinusoidal perturbation adaptive optimization. *Transactions of the Instrument Society of America*, **5,** No. 3, 255–262.

Lipovchak, M. S. (1964), EVOP at Chemcell. *Chemistry in Canada*, **16,** No. 3, 28–30.

Lowe, C. W. (1964), Some techniques of evolutionary operations. *Transactions of the Institution of Chemical Engineers*, **42,** T332–344.

McArthur, D. S. (1961), Strategy in research—alternative methods for design. *Institute of Radio Engineers, Transactions on Engineering Management*, **EM-8,** March, 34–40.

McArthur, D. S. and J. J. Heigl (1957), Strategy in research. Esso Research and Engineering Company.

Nelder, J. A., and R. Mead (1964), A simplex method for function minimization. *Computer Journal*, **7,** 308–313.

Newman, J. (1960), Examples of the practical use of response surface techniques and evolutionary operation. *EUCEPA*, IVth Symposium, Helsinki, May 16–18.

Pappas, M. A. (1962), EVOP requires communications to work. *Hydrocarbon Processing and Petroleum Refiner*, **41,** No. 5, 159–160.

Paris, W. W., and R. C. Manring (1961), Introduction to EVOP. Presented at the 139th ACS Meeting, St. Louis, Missouri, March 22.

Patnaik, P. B. (1949), The non-central χ^2- and F-distributions and their applications. *Biometrika*, **36,** 202–232.

Pearson, E. S., and H. O. Hartley (1954), *Biometrika Tables, Volume I.* Cambridge University Press, England. (Third edition, 1966.)

Pearson, E. S. and J. Wishart, editors (1958), "*Student's*" *Collected Papers*. Cambridge University Press, England, 181–191.

Pursglove, S. D. (1961), Here's how you can use EVOP. *Chemical Processing*, 23–24, 95–97 (August).

RAND Corporation (1955), *A Million Random Digits with 100,000 Normal Deviates*. The Free Press, Glencoe, Illinois.

Remmers, E. G. and C. G. Dunn (1961), Process improvement of a fermentation product. *Industrial and Engineering Chemistry*, **53,** No. 9, 743–745.

Riordan, F. S. (1958), Problems in EVOP programs. Presented at the Buffalo Conference, Chemical Division, ASQC, Buffalo, New York.

Riordan, F. S. (1959*a*), EVOP—revolution by evolution. Conference on Statistical Quality Control, co-sponsored by ASQC and Texas Manufacturers' Association.

Riordan, F. S. (1959*b*), Evolutionary operation. Paper presented at the Gordon Research Conference on Statistics, August 24.

Riordan, F. S. (1961), EVOP encapsulated. An unpublished paper.

Roberts, S. M., and T. M. Stout (1959), Problems and answers in the design of computer control systems. *Instruments and Control Systems*, **32,** 1855–1859.

Samuel, O. C. (1962), Experimenting on the plant process without interfering with production. *Food Processing*, **23,** No. 2, 31–33, 49.

Satterthwaite, F. E. (1959), Reply to discussion of papers by F. E. Satterthwaite and T. A. Budne on "Random Balance." *Technometrics*, **1,** 185.

Scheffé, H. (1959), *The Analysis of Variance*. Wiley, New York

Spendley, W., G. R. Hext, and F. R. Himsworth (1962), Sequential application of simplex designs in optimization and EVOP. *Technometrics*, **4,** 441–461.

Sweeney, R. F. et al. (1961), Mathematics, computers, operations research, and statistics. *Industrial and Engineering Chemistry*, **53,** No. 4, 329–336.

Thomas, M. D. (1965), Personal communication, December 13, to W. G. Hunter.

Thomas, M. D., and G. L. Webster (1960), Evolutionary operation applied to resistance welding of automotive sheet metal. An unpublished paper of General Motors Institute, Flint, Michigan.

Vajda, S. (1956), *The Theory of Games and Linear Programming*. Methuen, London.

Valer, A. (1966), Control of production process by evolutionary operation. *Chemicky prumysl*, **16,** No. 12, 705–709 (in Czechoslovakian).

Van Eck, L. F. (1958), Management's role in an EVOP program. Second Stevens Symposium Papers, 76, (January 26).

Van Eck, L. F. (1962), A path to more effective use of process data. *Industrial Quality Control*, **19,** No. 2, 8–10.

Wadehra, N. P. (1968), Design considerations of a combined brake-accelerator pedal system. M.S. Thesis, Kansas State University, Manhattan, Kansas.

Weaver, O. R. (1963), EVOP in operation. Paper read to the Seventh Technical Conference of the Chemical Division of ASQC, St. Paul, Minnesota, November 8.

Whitwell, J. C. (1959*a*), EVOP in chemical processes. *Tappi*, **42,** No. 6, 467–473.

Whitwell, J. C. (1959*b*), Practical applications of EVOP. *National Convention Transactions, ASQC*, 603–616.

Wilde, D. J. (1964), *Optimum Seeking Methods*. Prentice-Hall, Englewood Cliffs, New Jersey.

Wilson, C. L. (1960), A designed experiment. *Industrial and Engineering Chemistry*, **52,** No. 6, 504–506.

Youden, W. J. (1959), EVOP. *Industrial and Engineering Chemistry*, **51,** No. 6, 79A–80A.

Index

Across blocks, 133
Additivity of effects, 161
Aids to EVOP, 147
Alternative hypothesis, 55
Altpeter, R. J., 172, 194
American Cyanamid Company, 194
Area as probability, 38
Automatic EVOP, 193
Automatic optimization, 171, 193, 194
Auxiliary response, 189
Average, 42ff
Averaging, in EVOP, 49
 observations, 47

Batches, as blocks, 101
Batch operation, 23
Batch process, 207
Best-known conditions, 70
Between cycles estimate of σ, 77
Beyer, W. H., 89
Bibliography, 227ff
Bird-dog unit, 188
Block averages, 133, 134
Blocking, 150, 184
Blocking the 2^3 design, 88, 99ff
Block-to-block variation, 132ff
Box, G. E. P., 28, 74, 104, 109, 120, 156, 165, 167, 172, 184, 188, 192, 194, 200
British Museum, 9

Canadian Packers, Limited, 194
Canning industry, 194
Card game analogy, 190
Carryover, 27, 185
Celestial lottery, 32
Center of design, 102
Center point in 2^3 design, 77
Center points, 17
Central chi-square, 214
Central limit theorem, 36, 49, 51, 56
Cestoni, D. J., 194
Change in mean, 72–74, 102, 117
 2 S.E. limits, 104
Chanmugam, J., 172, 194
Characteristics of a distribution, 33
Charnes, A., 169
Checks, 129
Chemical engineer, 18, 147, 182
Chemstrand Corporation, 194, 195
Chess, 16
Chi-square variable, 213
Class frequency, 30, 31
Cochran, W. G., 74, 148, 167, 184
Comments on EVOP, 180ff
Companies mentioned, *see* individual names
Complicated processes, 180
Composite designs, 184
Computer, 105, 132, 182, 191
Confidence intervals, 55ff, 59
Confidence limits, 59, 62
Confounded, 100
Continuous process, 23, 207
Contours, 6, 189, 198
 by eye, 78, 94
Contrasts, 50ff, 67, 88
Control, 175, 182, 192
Cooper, W. W., 169
Correction for the mean, 43
Correlation, 25ff, 103, 185, 192
Cost of EVOP, 73–74

Coutie, G. A., 194
Covariance, 53
Cox, G. M., 74, 148, 167, 184
Cube, diagonal planes, 101
faces, 101
Current best-known conditions, 70
Cycle, 14, 107

David, H. A., 134
Davies, O. L., 4, 61, 74, 148, 165, 167, 184, 218–220
De Busk, R. E., 194, 195
Decisions in EVOP, 15
Design, 13, 150
factorial, 63ff
patterns, 184
simplex, 178
Detectability of effects, 211
Diagonal planes of cube, 101
Discrete valued responses, 18
Distribution of sample average, 46
Distributions, 30ff
Dot diagram, 23, 29
Dow Chemical Company, 194
Draper, C. S., 172
Draper, N. R., 192

Educational aspects of EVOP, 150
Effect, 12, 51, 150, 201
interpretation of, 69ff
Effect distribution, 201
Efficiency of estimates, 45, 99
Electronic computer, 105, 132, 182, 191
Empirical feedback, 153, 156ff, 159, 173, 179, 193, 200ff
Error limits, 14
of first kind, 211
of second kind, 211
Estimate, of mean, 41ff
of standard deviation, 41ff, 94, 132, 196
Evolution of lobsters, 9
EVOP, committee, 18, 147, 182, 190
course, 136, 151
game, 141, 198
pronunciation, 5
Eye-drawn contours, 78, 94

Faces of cube, 101
Factor, $f_{4,n}$, 117, 132
$f_{5,n}$, 115, 132
$f_{k,n}$, 97, 197
$f_{k,n}$ table, 98, 222

Factorial designs, 63
Failures in EVOP, 149
Feedback, empirical, 153, 200ff
scientific, 19, 153ff, 164, 179, 200, 208, 209
Felice, B., 194
Fibonacci search, 175
Fisher, R. A., 220
Fix, E., 214
Frequency distributions, 30

Game, 141, 198

Hadley, G., 160, 169
Harrington, E. C., 175, 176
Hartley, H. O., 218, 219, 222
Hehner, H. O., 194, 195
Height of the grass, 12
Henderson, A., 169
Hext, G. R., 176, 178
High variability, 181
Himsworth, F. R., 176, 178
How many cycles?, 211
Hunter, J. S., 109, 120, 184
Hunter, W. G., 194

Imperial Chemical Industries, Limited, 194
Important variables, 180
Independence of observations, 28
Independent variables, 52, 53
Indicator unit, 188
Information board, 10, 13, 15, 77, 92, 107, 190
EVOP game, 142ff
sputnik, 138
Interaction, 68, 150, 184
in 2^2, 67 (figure), 68ff
in 2^3, 84ff, 85 (figure)
Interpolation for contours, 78
Interpretation of effects, in 2^2, 69ff
Iterative nature of experimentation, 152
Jenkins, G. M., 28, 185

Keeping, E. S., 213
Kenworthy, I. C., 178
Kittrell, J. R., 194
Klingel, A. R., 194
Koehler, T. L., 194
Koleff, A. M., 194
Kotnour, K. D., 172, 194

Lag 1 serial correlation, 25ff
Latent variable, 192
Least-squares, 191, 192
Level of significance, 56, 57
Li, Y. T., 172
Linear combination, 51
Linear programming, 160, 168ff, 189
Lobster, evolution of, 9
Location, 30, 33
London Sunday Express, 9
Lottery, 32, 46
Lowe, C. W., 176
Lurking variable, 192

MacIntyre, R. G., 194
Main effects, for 2^2, 66ff, 67 (figure)
for 2^3, 80ff, 82 (figure)
Management, 135ff
Mathematician, 182
Maumee Chemical Company, 194
Mean, 33
of contrasts, 50ff
of difference, 53
of sum, 53
Models for empirical feedback, 156ff
Modifications of EVOP, 175
Monsanto Company, 139, 175, 194, 195
Multiple regression, 191
Multiple responses, 150, 162, 170, 188

Natural selection, 8
Negative serial correlation, 26, 27, 185
New EVOP pattern, 146
New ideas in EVOP, 149
Neyman-Pearson theory, 211
Noise level, 12
Noncentral chi-square, 213, 214
Normal deviates, 60, 142, 199, 224–226 (table)
Normal distribution, 35ff, 198, 199, 210, 213, 218–221 (table)
Null hypothesis, 55, 57
Number of cycles, 211
Number of variables, 183

Objections to EVOP, 180ff
Objective of EVOP game, 147
Observations, 23
Olson, L. R., 194
One factor at a time design, 83
One-sided probabilities, 39
Operators, 137ff
Optimal conditions, 161
Optimal empirical feedback, 200ff
Optimization, 152, 178
automatic, 171, 193, 194
methods, 160ff
techniques related to EVOP, 164ff
techniques unrelated to EVOP, 175
Overall change in mean, 104
Organization of EVOP, 135
Orthogonality, 101

Pappas, M. A., 194
Parallel units, 188
Partan, 175
Past data in EVOP, 184
Patnaik, P. B., 214
Patterns of variants, 184
Pearson, E. S., 218, 219, 222
Phase, 13, 15, 107
mean, 72, 104, 132, 133
Phillips Petroleum Company, 194
Plotting data in time order, 24
Point of inflexion, 38
Population, 31ff
Positive serial correlation, 25, 26, 185
Preferable conditions, 161
Principal response, 189
Prior estimate of σ, 108
Probability, as area, 38
points of normal distribution, 218–221
Probability density, 31
Probability distribution, 31ff
Probability statement, one- or two-sided, 39
Process control, 182
Process improvement, 182
Process modeling, 182
Process modification, 200
Process superintendent, 18, 20, 174, 190
Profit, 21, 203, 207
Pronunciation of EVOP, 5
Pursglove, S. D., 194

Qualitative variables, 63
Quality control, 182
Quality control man, 147, 148
Quantative variables, 63
Queries on EVOP, 180

Random drawing, 32
Randomization, 75

Random normal deviates, 60, 142, 199, 224–226 (table)
Random order, 188
Random sample, 32
Random search, 175
Random variables, 51
Random variation, 28
Range, 44, 196
 estimate of σ, 44ff
 factor, 45 (and table), 95, 96, 222 (table)
Recognition of success in EVOP, 149
Reference conditions, 71, 102, 108, 184
References, 227ff
Regular simplex, 177
Relative efficiency, 45
Replication, 178
Research man, 18, 147, 148
Residuals, 43
Resisting statisticians, 148
Response curve drift, 171
Response surface, 6, 166, 190, 194, 198, 199
REVOP, 176
Ringelman, R. E., 194
Riordan, F. S., 194, 195
ROVOP, 175
Run, 23, 107, 150
Runs sequence, 75, 103
Runs standard order, 79

Sample, 31–33, 41ff
 average or mean, 42
 variance, 42ff
Sampling distribution of average, 46
Samuel, O. C., 194
Satterthwaite, F. E., 176
Savings, 21, 195
Scaleup, 6, 190
Scaling of variables, 164
Scientific feedback, 19, 153ff, 164, 179, 200, 208, 209
Screening, 164, 175
Second-order designs, 184
Selection, 10
Sequence numbering of runs, 75, 103
Sequential significance test, 57ff
Serial correlation, 25ff, 103, 185, 188
"Short-cut" formula for variance, 43
Significance level, 56, 57
Significance tests, 55ff, 211
 sequential, 57ff
Simplex, regular, 177
Simplex design, 178
Simplex EVOP, 176, 209
Simulation, 182
 of EVOP, 136, 141ff, 198ff
Skewed distribution, 49
Small-scale studies, 190
Smith, H., 192
Specialists, 20
Specifications, 181
Spendley, W., 176, 178
Spread, 30
 standard deviation, 34
Sputnik, 138
Staley (A.E.) Manufacturing Company, 194
Standard deviation, 12, 34
 of block averages, 133
 calculation, 107, 128
 estimate, 94, 196
Standard error of phase mean, 132ff
Standard errors, of contrasts, 54
 of effects, 75ff
 of effects for 2^p, 91ff
Standard Oil of Ohio, 194
Standard run order, 79, 90
Static operation, 10, 191
Statistical independence, 28, 52
Statistically significant, 55, 207
Statistician, 19, 147, 148, 168, 190
Steepest ascent, 165
Sum of squares check, 90
Sunday Express, 9
Supercats, 9
Supervisory personnel, 136ff
Surface contours, 6, 7
Swift and Company, 194

Tables, 217ff
t distribution, 60
Tennessee Eastman Co., 194, 195
Tests of significance, 55ff, 211
Thomas, M. D., 195
Three-factor interaction in 2^3, 86, 87 (figure)
Three-variable EVOP, 16ff, 119ff (worksheets)
Time order plot of data, 24
Time periods as blocks, 101
Time trends, 192
Training programs, 135ff
Transformations, 190
Tuning, 4, 7

2^2 design, 63, 65ff
with center point, 77
effects, 66ff
2^3 design, 63, 79ff
blocking, 99ff
summary, 88
2^p design, 184
Two-factor interactions, in 2^2, 68ff
in 2^3, 84ff
2 S.E. limits, 14, 61, 113, 128, 189
for change in mean, 104
Two-sided probabilities, 39
Two-variable EVOP, 13ff, 105ff (worksheets)
Two-way analysis of variance, 188
Two-way table in 2^3, 93

Uncorrected sum of squares, 43

Vajda, S., 169
Variance, 33ff
of contrasts, 50ff
of difference, 53
of effects in 2^p, 91
of sum, 53
Variation, 10
in results, 24
Variations of EVOP, 152

Watts, D. G., 28, 185
Weaver, O. R., 194
Webster, G. L., 195
Wiener process, 201
Wilde, D. J., 175
Wilson, C. L., 194
Wilson, K. B., 165, 167
Within blocks, 133, 134
Within cycles estimate of σ, 77
Worksheets, for 2^2 design, 105ff
for 2^3 design, 119ff
Works process, 10, 12, 15–17

Yates, F., 220
Yates' algorithm, 89ff, 118, 128 ,